STUDYING ARCTIC FIELDS

McGILL-QUEEN'S NATIVE AND NORTHERN SERIES
(In memory of Bruce G. Trigger)
Sarah Carter and Arthur J. Ray, Editors

1 When the Whalers Were Up North
Inuit Memories from the Eastern Arctic
Dorothy Harley Eber

2 The Challenge of Arctic Shipping
Science, Environmental Assessment, and Human Values
Edited by David L. VanderZwaag and Cynthia Lamson

3 Lost Harvests
Prairie Indian Reserve Farmers and Government Policy
Sarah Carter

4 Native Liberty, Crown Sovereignty
The Existing Aboriginal Right of Self-Government in Canada
Bruce Clark

5 Unravelling the Franklin Mystery
Inuit Testimony
David C. Woodman

6 Otter Skins, Boston Ships, and China Goods
The Maritime Fur Trade of the Northwest Coast, 1785–1841
James R. Gibson

7 From Wooden Ploughs to Welfare
The Story of the Western Reserves
Helen Buckley

8 In Business for Ourselves
Northern Entrepreneurs
Wanda A. Wuttunee

9 For an Amerindian Autohistory
An Essay on the Foundations of a Social Ethic
Georges E. Sioui

10 Strangers Among Us
David Woodman

11 When the North Was Red
Aboriginal Education in Soviet Siberia
Dennis A. Bartels and Alice L. Bartels

12 From Talking Chiefs to a Native Corporate Elite
The Birth of Class and Nationalism among Canadian Inuit
Marybelle Mitchell

13 Cold Comfort
My Love Affair with the Arctic
Graham W. Rowley

14 The True Spirit and Original Intent of Treaty 7
Treaty 7 Elders and Tribal Council with Walter Hildebrandt, Dorothy First Rider, and Sarah Carter

15 This Distant and Unsurveyed Country
A Woman's Winter at Baffin Island, 1857–1858
W. Gillies Ross

16 Images of Justice
Dorothy Harley Eber

17 Capturing Women
The Manipulation of Cultural Imagery in Canada's Prairie West
Sarah Carter

18 Social and Environmental Impacts of the James Bay Hydroelectric Project
Edited by James F. Hornig

19 Saqiyuq
Stories from the Lives of Three Inuit Women
Nancy Wachowich in collaboration with Apphia Agalakti Awa, Rhoda Kaukjak Katsak, and Sandra Pikujak Katsak

20 Justice in Paradise
Bruce Clark

21 Aboriginal Rights and Self-Government
The Canadian and Mexican Experience in North American Perspective
Edited by Curtis Cook and Juan D. Lindau

22 Harvest of Souls
The Jesuit Missions and Colonialism in North America, 1632–1650
Carole Blackburn

23 Bounty and Benevolence
A History of Saskatchewan Treaties
Arthur J. Ray, Jim Miller, and Frank Tough

24 The People of Denendeh
Ethnohistory of the Indians of Canada's Northwest Territories
June Helm

25 The *Marshall* Decision and Native Rights
Ken Coates

26 The Flying Tiger
Women Shamans and Storytellers of the Amur
Kira Van Deusen

27 Alone in Silence
European Women in the Canadian North before 1940
Barbara E. Kelcey

28 The Arctic Voyages of Martin Frobisher
An Elizabethan Adventure
Robert McGhee

29 Northern Experience and the Myths of Canadian Culture
Renée Hulan

30 The White Man's Gonna Getcha
The Colonial Challenge to the Crees in Quebec
Toby Morantz

31 The Heavens Are Changing
Nineteenth-Century Protestant Missions and Tsimshian Christianity
Susan Neylan

32 Arctic Migrants/Arctic Villagers
The Transformation of Inuit Settlement in the Central Arctic
David Damas

33 Arctic Justice
On Trial for Murder – Pond Inlet, 1923
Shelagh D. Grant

34 The American Empire and the Fourth World
Anthony J. Hall

35 Eighteenth-Century Naturalists of Hudson Bay
Stuart Houston, Tim Ball, and Mary Houston

36 Uqalurait
An Oral History of Nunavut
Compiled and edited by John Bennett and Susan Rowley

37 Living Rhythms
Lessons in Aboriginal Economic Resilience and Vision
Wanda Wuttunee

38 The Making of an Explorer
George Hubert Wilkins and the Canadian Arctic Expedition, 1913–1916
Stuart E. Jenness

39 Chee Chee
A Study of Aboriginal Suicide
Alvin Evans

40 Strange Things Done
Murder in Yukon History
Ken S. Coates and William R. Morrison

41 Healing through Art
Ritualized Space and Cree Identity
Nadia Ferrara

42 Coyote and Raven Go Canoeing
Coming Home to the Village
Peter Cole

43 Something New in the Air
The Story of First Peoples Television Broadcasting in Canada
Lorna Roth

44 Listening to Old Woman Speak
Natives and Alternatives in Canadian Literature
Laura Smyth Groening

45 Robert and Francis Flaherty
A Documentary Life, 1883–1922
Robert J. Christopher

46 Talking in Context
Language and Identity in Kwakwaka'wakw Society
Anne Marie Goodfellow

47 Tecumseh's Bones
Guy St-Denis

48 Constructing Colonial Discourse
Captain Cook at Nootka Sound
Noel Elizabeth Currie

49 The Hollow Tree
Fighting Addiction with Traditional Healing
Herb Nabigon

50 The Return of Caribou to Ungava
A.T. Bergerud, Stuart Luttich, and Lodewijk Camps

51 Firekeepers of the Twenty-First Century
First Nations Women Chiefs
Cora J. Voyageur

52 Isuma
Inuit Video Art
Michael Robert Evans

53 Outside Looking In
Viewing First Nations Peoples in Canadian Dramatic Television Series
Mary Jane Miller

54 Kiviuq
An Inuit Hero and His Siberian Cousins
Kira Van Deusen

55 Native Peoples and Water Rights
Irrigation, Dams, and the Law in Western Canada
Kenichi Matsui

56 The Rediscovered Self
Indigenous Identity and Cultural Justice
Ronald Niezen

57 As affecting the fate of my absent husband
Selected Letters of Lady Franklin Concerning the Search for the Lost Franklin Expedition, 1848–1860
Edited by Erika Behrisch Elce

58 The Language of the Inuit
Syntax, Semantics, and Society in the Arctic
Louis-Jacques Dorais

59 Inuit Shamanism and Christianity
Transitions and Transformations in the Twentieth Century
Frédéric B. Laugrand and Jarich G. Oosten

60 No Place for Fairness
Indigenous Land Rights and Policy in the Bear Island Case and Beyond
David T. McNab

61 Aleut Identities
Tradition and Modernity in an Indigenous Fishery
Katherine L. Reedy-Maschner

62 Earth into Property
Aboriginal History and the Making of Global Capitalism
Anthony J. Hall

63 Collections and Objections
Aboriginal Material Culture in Southern Ontario, 1791–1914
Michelle A. Hamilton

64 These Mysterious People
Shaping History and Archaeology in a Northwest Coast Community, Second Edition
Susan Roy

65 Telling It to the Judge
Taking Native History to Court
Arthur J. Ray

66 Aboriginal Music in Contemporary Canada
Echoes and Exchanges
Edited by Anna Hoefnagels and Beverley Diamond

67 In Twilight and in Dawn
A Biography of Diamond Jenness
Barnett Richling

68 Women's Work, Women's Art
Nineteenth-Century Northern Athapaskan Clothing
Judy Thompson

69 Warriors of the Plains
The Arts of Plains Indian Warfare
Max Carocci

70 Reclaiming Indigenous Planning
Edited by Ryan Walker, Ted Jojola, and David Natcher

71 Setting All the Captives Free
Capture, Adjustment, and Recollection in Allegheny Country
Ian K. Steele

72 Before Ontario
The Archaeology of a Province
Edited by Marit K. Munson and Susan M. Jamieson

73 Becoming Inummarik
Men's Lives in an Inuit Community
Peter Collings

74 Ancient Pathways, Ancestral Knowledge
Ethnobotany and Ecological Wisdom of Indigenous Peoples of Northwestern North America
Nancy J. Turner

75 Our Ice Is Vanishing/Sikuvut Nunguliqtuq
A History of Inuit, Newcomers, and Climate Change
Shelley Wright

76 Maps and Memes
Redrawing Culture, Place, and Identity in Indigenous Communities
Gwilym Lucas Eades

77 Encounters
An Anthropological History of Southeastern Labrador
John C. Kennedy

78 Keeping Promises
The Royal Proclamation of 1763, Aboriginal Rights, and Treaties in Canada
Edited by Terry Fenge and Jim Aldridge

79 Together We Survive
Ethnographic Intuitions, Friendships, and Conversations
Edited by John S. Long and Jennifer S.H. Brown

80 Canada's Residential Schools
The History, Part 1, Origins to 1939
The Final Report of the Truth and Reconciliation Commission of Canada, Volume 1

81 Canada's Residential Schools: The History, Part 2, 1939 to 2000
The Final Report of the Truth and Reconciliation Commission of Canada, Volume 1

82 Canada's Residential Schools: The Inuit and Northern Experience
The Final Report of the Truth and Reconciliation Commission of Canada, Volume 2

83 Canada's Residential Schools: The Métis Experience
The Final Report of the Truth and Reconciliation Commission of Canada, Volume 3

84 Canada's Residential Schools: Missing Children and Unmarked Burials
The Final Report of the Truth and Reconciliation Commission of Canada, Volume 4

85 Canada's Residential Schools: The Legacy
The Final Report of the Truth and Reconciliation Commission of Canada, Volume 5

86 Canada's Residential Schools: Reconciliation
The Final Report of the Truth and Reconciliation Commission of Canada, Volume 6

87 Aboriginal Rights Claims and the Making and Remaking of History
Arthur J. Ray

88 Abenaki Daring
The Life and Writings of Noel Annance, 1792–1869
Jean Barman

89 Trickster Chases the Tale of Education
Sylvia Moore

90 Secwépemc People, Land, and Laws
Yerí7 re Stsq́eýs-kucw
Marianne Ignace and Ronald E. Ignace

91 Travellers through Empire
Indigenous Voyages from Early Canada
Cecilia Morgan

92 Studying Arctic Fields
Cultures, Practices, and Environmental Sciences
Richard C. Powell

STUDYING ARCTIC FIELDS

CULTURES, PRACTICES, AND ENVIRONMENTAL SCIENCES

RICHARD C. POWELL

McGill-Queen's University Press
Montreal & Kingston | London | Chicago

ISBN 978-0-7735-5112-1 (cloth)
ISBN 978-0-7735-5113-8 (paper)
ISBN 978-0-7735-5255-5 (ePDF)
ISBN 978-0-7735-5256-2 (ePUB)

Legal deposit fourth quarter 2017
Bibliothèque nationale du Québec

Printed in Canada on acid-free paper that is 100% ancient forest free (100% post-consumer recycled), processed chlorine free

Funding for this project has been provided by the Canada-UK Foundation.

McGill-Queen's University Press acknowledges the support of the Canada Council for the Arts for our publishing program. We also acknowledge the financial support of the Government of Canada through the Canada Book Fund for our publishing activities.

Library and Archives Canada Cataloguing in Publication

Powell, Richard C., author
Studying Arctic fields : cultures, practices, and environmental sciences / Richard C. Powell.

(McGill-Queen's Native and northern series ; 92)
Includes bibliographical references and index.
Issued in print and electronic formats.
ISBN 978-0-7735-5112-1 (cloth). – ISBN 978-0-7735-5113-8 (paper). – ISBN 978-0-7735-5255-5 (ePDF). – ISBN 978-0-7735-5256-2 (ePUB)

1. Arctic regions – Research – Canada. 2. Research – Social aspects – Canada. 3. Research – Political aspects – Canada. 4. Environmental sciences – Canada. 5. Polar Continental Shelf Project (Canada). I. Title. II. Series: McGill-Queen's Native and northern series ; 92

Q180.C3P69 2018 507.2'09719 C2017-904836-8
C2017-904837-6

Set in 11/13.5 Minion Pro with Avenir Next
Book design & typesetting by Garet Markvoort, zijn digital

FOR JO, BETSAN, ALYS, AND HANA

CONTENTS

Figures | xiii

Prologue | xv

Acknowledgments | xvii

Introduction | 3

1 Scientific Sovereignty, Nordicity, and the Canadian Nation | 27

2 Between Observation and Experiment in Arctic Fieldwork | 54

3 Base Cultures: The Spatial Organization of a Research Station | 76

4 Performing the Arctic Scientific Human | 98

5 Canada Day in Qausuittuq: Dramatizing Inuit Encounters | 125

6 Emotional Practices and Play: The Quotidian Provenance of Logistics | 150

7 Hidden Voices? Competing Visions and the Everyday Governance of Arctic Science | 163

Epilogue: Requiem for a *Canadian* Arctic? | 188

Note on Methodology, Sources, and Research Ethics | 193
Notes | 197
References | 215
Index | 233

FIGURES

All photographs by author unless otherwise stated.

0.1 Main building, PCSP Resolute, August 2001 | 4
0.2 Places mentioned in text | 5
0.3 Canadian Arctic Islands, indicating JAWS and Inuit communities | 10
0.4 Resolute Airport Terminal, July 2002 | 11
1.1 PCSP aerial photograph of Ward Hunt Island, 13 June 1962. © Government of Canada. Reproduced with the permission of Library and Archives Canada (2016). Source: Library and Archives Canada/Natural Resources Canada fonds/PA-213932 | 45
1.2 PCSP Arctic Field Operations, 1959–1966. Redrawn from original map, "Polar Continental Shelf Project, Area of Arctic Field Operations, 1959–1966," LAC, RG45, vol. 336, F1-4 (pt 2) Expenditure Plans, Expenditure Votes, Program Plans and Related Records (1965–66) | 47
1.3 PCSP survey party at "hill" of granitic boulders on Arctic Ocean, 20 km northwest of Cape Isachsen, Ellef Ringnes Island, 24 April 1959. © Government of Canada. Reproduced with the permission of Library and Archives Canada (2016). Source: Library and Archives Canada/Natural Resources Canada fonds/EFR-59-8-1 | 50
2.1 Fred Roots making a presentation, circa 1961. © Government of Canada. Reproduced with the permission of Library and Archives Canada (2016). Source: Library and Archives Canada/Hans Walter Pulkkinen fonds/PA213942 | 63
2.2 Helicopter landing on an ice ridge, May 1971. © Government of Canada. Reproduced with the permission of Library and Archives

Canada (2016). Source: Library and Archives Canada/Hans Walter Pulkkinen fonds/PA213948 | 65
2.3 F.P. Hunt using a Tellurometer in field near Camp 200, 1970. © Government of Canada. Reproduced with the permission of Library and Archives Canada (2016). Source: Library and Archives Canada/Hans Walter Pulkkinen fonds/PA213947 | 66
2.4 Prince Patrick, Ellef Ringnes, and Brock Islands | 72
2.5 PCSP Ice Patrol, "Multi-year ice near Camp 200," 1970. © Government of Canada. Reproduced with the permission of Library and Archives Canada (2016). Source: Library and Archives Canada/Natural Resources Canada fonds/PA-213895 | 73
3.1 Base office, PCSP Resolute, outside, August 2001 | 80
3.2 Interior of base office, PCSP Resolute, August 2001 | 81
3.3 High Arctic fly-camps, base office, PCSP Resolute, August 2001 | 81
3.4 Twin Otter, PCSP Resolute, August 2001 | 83
3.5 Helicopter, PCSP Resolute, August 2001 | 83
3.6 Hangar, PCSP Resolute, August 2001 | 91
3.7 Dry laboratory, PCSP Resolute, August 2001 | 92
3.8 Library, PCSP Resolute, August 2001 | 95
3.9 Kitchen and dining hall, PCSP Resolute, August 2001 | 96
4.1 Botanists in the field, near Resolute, August 2001 | 113
4.2 Hydrologist performing measurements in the field, near Resolute, June 2002 | 115
4.3 Hydrologist in wetland, near Resolute, July 2002 | 116
4.4 Hydrologist using ATV to move between field sites, near Resolute, July 2002 | 117
4.5 Parks Canada base, Tanquary Fiord, Quttinirpaaq (Ellesmere Island National Park), July 2002 | 119
5.1 Hamlet of Resolute Bay, August 2001 | 132
5.2 Vehicle parade on Canada Day passing Polar Continental Shelf Project base complex, July 2002 | 141
5.3 Community celebrations on Canada Day, near Resolute, July 2002 | 144
5.4 Community river-crossing races, near Resolute, July 2002 | 145
5.5 Canadian flag displayed during Canada Day celebrations, July 2002 | 146

PROLOGUE

By the late hours of 19 October 2015, it was evident that a remarkable political event had occurred in Canada. Against all previous polling, the Liberal Party led by Justin Trudeau had taken a huge majority in the forty-second parliamentary general election. After nearly a decade of governance by Prime Minister Stephen Harper of the Conservative Party of Canada, Canadians thoroughly rejected his campaign vision. Justin, the eldest son of the erstwhile Liberal prime minister Pierre Trudeau, had returned his party to power in the most remarkable circumstances. Having held just thirty-six seats in the previous Parliament, the Liberal Party took 184 ridings of 338, with huge regional gains across Canada.

The 2015 election was seen by many commentators as a referendum on the Harper government, because of the changes that had been initiated during his administration. For Harper, Canada was a strong country, a country that developed its resources and defended its borders to ensure economic growth. Harper had made the development of the Arctic a key plank of his domestic policy, and had made Arctic sovereignty central to his foreign policy. Further, the Harper administration retrenched public servants and restricted the ability of federal scientists to talk to the media, lest they dissent from government policy. The sometime canonical principles of Canadian national identity concerning environmental stewardship, peacekeeping, bilingualism, and the accommodation of indigenous rights and "new Canadians" were, according to Harper, just myths. Indeed, they were *Liberal* myths. Canadians built a post-1945 national identity on the idea that they were not Americans. Harper, ultimately, thought that this had made Canada weak. By superseding the United States in natural resource development and financial deregulation, Harper aimed to make Canada strong. As his 2015 election campaign had it, the Conservative Party offered "Proven leadership for a strong Canada."

A key dimension of Harper's policy for northern Canada was the 2010 announcement of the building of the Canadian High Arctic Research Station (CHARS) in Cambridge Bay, Nunavut. Although construction did not actually commence until 2014, the CHARS was to be the centre for internationally excellent scientific research into natural resource development in the Arctic region. Harper envisaged the CHARS, scheduled to open in 2017, as consolidating his vision for a particular type of *scientific sovereignty* in the Canadian Arctic.

In opposition, the younger Trudeau articulated a different view of science and politics in Canada. Despite Harper's emphasis on the Arctic as a policy objective, the entire Canadian Arctic of Nunavut, the Yukon, and the Northwest Territories elected Liberal representatives. In the words of Trudeau's victory speech in Montreal, late into the night of 19 October, "A positive, optimistic, hopeful vision of public life isn't a naive dream. It can be a powerful force for change ... I will be the prime minister of all Canadians ... This is Canada, and in Canada, 'Better Is Always Possible.'" By March 2016, the Liberal government had announced that the mandate for the CHARS was to change to focus on climate change in the northern regions.

Why does this concern us? This book is about the relations between science, politics, and people in the Canadian Arctic. From the earliest stages of my research, I was warned that Canada's Polar Continental Shelf Project had a distinctive *culture of field practice*. This book tries to explain why that might be. The answer, it seems, is partly about site and location, and about how science is practised in difficult environments. But the answer is also about the conduct of science as public service. It is about what it means to be Canadian.

ACKNOWLEDGMENTS

Like most books, this one is the outcome of an intellectual odyssey that has encompassed various sites and taken too many years. It has relied on the encouragement of many, and on the sympathy of many more.

During the earliest beginnings of this project as a Canadian Rhodes Scholars Foundation Scholar at the University of British Columbia, I was exposed to a vast range of social and critical theory and to the fascinating histories and geographies of Canada. I must thank a number of people for encouraging me to follow my interests back then: Cole Harris, Mike Church, Derek Gregory, Bob MacDonald, Mark Phillips, Alan Richardson, Judy Segal, Olav Slaymaker, and Graeme Wynn. The friendship and support of Trevor Barnes was critical in keeping going, and John Stager's help was crucial as I was developing my interest in the Canadian north. Friends from those Vancouver days number Simon Dadson, Matt Farish, Graham Horner, Arn Keeling, Anthony Story, Alex Vasudevan, Johanna Waters, Bob Wilson, and Jamie Winders, among many others. The Department of Geography and Scott Polar Research Institute at the University of Cambridge then provided an excellent site from which to conduct much of my initial Arctic fieldwork. For support and conversations in the spaces in between, I am grateful to Michael Bravo, Keith Richards, Tim Bayliss-Smith, Laura Cameron, Gerry Kearns, Nick Megoran, Heike Jöns, David Livingstone, and Peter Meusburger. In Manchester, I am immensely grateful for the help and support of Noel Castree and Chris Perkins. During the later research conducted in Liverpool, I benefited from conversations with Richard Phillips, Dave Featherstone, Bethan Evans, Clare Holdsworth,

Andy Davies, and Tinho Da Cruz. The late Bob Woods was a constant source of encouragement. In the process of bringing things together in Oxford, I have enjoyed the company of Andrew Barry, Judy Pallot, Linda McDowell, Derek McCormack, Pam Berry, Tony Lemon, Thomas Jellis, Georgina Born, Sarah Whatmore, and the fellowship of Mansfield College. I have recently been appointed to a position back in Cambridge and look forward to rejoining colleagues there.

The project has been expensive and thus required funding from numerous sources. Some of the work draws upon doctoral research undertaken through an ESRC/NERC Interdisciplinary Research Studentship (R42200034029). Other generous support has been provided by an ESRC/RCEP Interdisciplinary Research Fellowship (RES-152-27-0002), an ESRC Postdoctoral Fellowship (PTA-026-27-0112), a Simon Research Fellowship (University of Manchester), Emmanuel College, Cambridge (External Research Award, Sansom Fund Award, and Bachelor Scholarship Fund for Fieldwork Award), the University of Cambridge (two Smuts Memorial Fund Awards and a Bartle Frere Exhibition), the Department of Geography, University of Cambridge (Philip Lake Fund Award), the History and Philosophy of Geography Research Group, Royal Geographical Society with the Institute of British Geographers (two Postgraduate Awards), the Historical Geography Research Group, Royal Geographical Society with the Institute of British Geographers (Postgraduate Research Award), the Royal Society, London (two Dudley Stamp Memorial Trust Awards), the International Council for Canadian Studies, Ottawa (Graduate Student Scholarship), the University of Ottawa (Canadian Studies Institutional Support Award), the Canadian Association of Geographers (Student Travel Award), and the British Association for Canadian Studies with the Foundation for Canadian Studies in the UK (Molson Research Award).

Philip Cercone, Kathleen Fraser, and the staff at McGill-Queen's University Press have been wonderful to work with, and I am particularly grateful to the reviewers for their comments on versions of this manuscript. Some passages in the text draw from my earlier papers and are rewritten and extended with grateful permission here. Sections of chapter 1 draw upon "Science, Sovereignty and Nation: Canada and the Legacy of the International Geophysical Year, 1957–1958" (2008c). Some passages in chapter 2 appeared in an earlier version in "'The

Rigours of an Arctic Experiment': The Precarious Authority of Field Practices in the Canadian High Arctic" (2007b). A shorter version of chapter 5 was published as "Canada Day in Resolute: Performance, Ritual and the Nation in an Inuit Community" (2009a). Some sections of chapter 6 are rewritten from "Learning from Spaces of Play: Recording Emotional Practices in High Arctic Environmental Sciences" (2009b).

I cannot possibly thank by name all those individuals across Canada who participated in this project, by opening themselves for investigation or simply passing on documents and information, but I must mention a few. In Ottawa, the research would have been impossible without the help and friendship of Susan Aiken and Peter Johnson, and the time and kindness of the sometime directors of the Polar Continental Shelf Project, Fred Roots, George Hobson, and Bonni Hrycyk. At the National Archives of Canada, I am indebted to Doug White and Mark Levene for help, and to Rocky Crupi, Tony Bonnaci, and Nathalie Dussault for promptly reviewing seemingly endless requests for documents under great constraints.

In Resolute, I am immensely grateful to those who allowed me to observe and collaborate with them in the field. Thank you, too, to all the PCSP base staff and scientists, your patient toleration of my constant presence and annoying questions is evident in what follows. Thanks also to everyone at Tanquary Fiord on Ellesmere. I have undertaken not to name any of you, and to disguise your identities, but I hope I have gone some way to both relaying your stories and capturing something of their importance.

Thank you again to my mother for her unfailing support of my education. Most of all, thank you to my wife, Jo, who has been an intellectual interlocutor in this project and companion in all others. Our daughters, Betsan Cheryl, Alys Seren, and Hana Carys, have put all these long years of study into perspective.

STUDYING ARCTIC FIELDS

INTRODUCTION

> The Arctic climate system is characterized by its low thermal energy state and intimate couplings between the atmosphere, ocean and land. This makes the Arctic a challenging and fascinating region to study. But given the complex nature of the Arctic climate system, where does one start in describing it?
>
> Mark Serreze and Roger Barry, 2005, *The Arctic Climate System*, 17

> Dans le Grand Nord, l'espace absorbe le temps et le matérialise en étendues sublimes.
>
> Michel Onfray, 2002, *Esthétique du Pôle Nord*, 49

In recent years, the changing environments of the Arctic have attracted much commentary. It is widely appreciated that the Circumpolar Region is critical to understanding global climate change. Such discourses, in the public mind at least, seem finally to have replaced the recurrent notion of the Arctic Sublime. The impacts of these environmental shifts upon associated social worlds, especially those of the Inuit who dwell in the high latitudes, have been granted some, if often cursory, attention. But the social worlds of those scientists who spend much of their working lives *investigating* those environmental changes have not. Such Arctic scientists are the very people who have attempted to take the messages of local changes in ice behaviour or caribou migration to global audiences. As recent work by ethnographers such as Phaedra Daipha (2015) argues, climate scientists believe that they hold the key to the most pressing questions facing the future of humanity, and those undertaking fieldwork in the Arctic scrutinize these processes at their vanguard. This book is about the social and cultural lives

0.1 | Main building, PCSP Resolute, August 2001

of those individuals. What happens in those remote, often isolated sites on the globe where questions about planetary futures are investigated? Too often, when environmental futures are envisaged as at stake, these questions are waved away as lacking importance. On the contrary, as we will discover, the social worlds of Arctic environmental scientists impinge on the very structuring of research questions and the "delivery technologies" of their answers. Indeed, without the cultural life of Arctic fieldwork, there would not be any scientific practice. Without such practices, the Circumpolar Arctic would not have become *the* signifying region of global climate change. In what follows, I open up these practices to analysis and delineate their expanding social lives.

This book, then, is an ethnography of field science.[1] It depicts what actually happens in the quotidian production of scientific activity in the Arctic. In part, it is a study of a particular place – a field station at Resolute, Nunavut, in the Canadian High Arctic (figure 0.1). At the same time, it is an examination of the growth of a project, a mode of

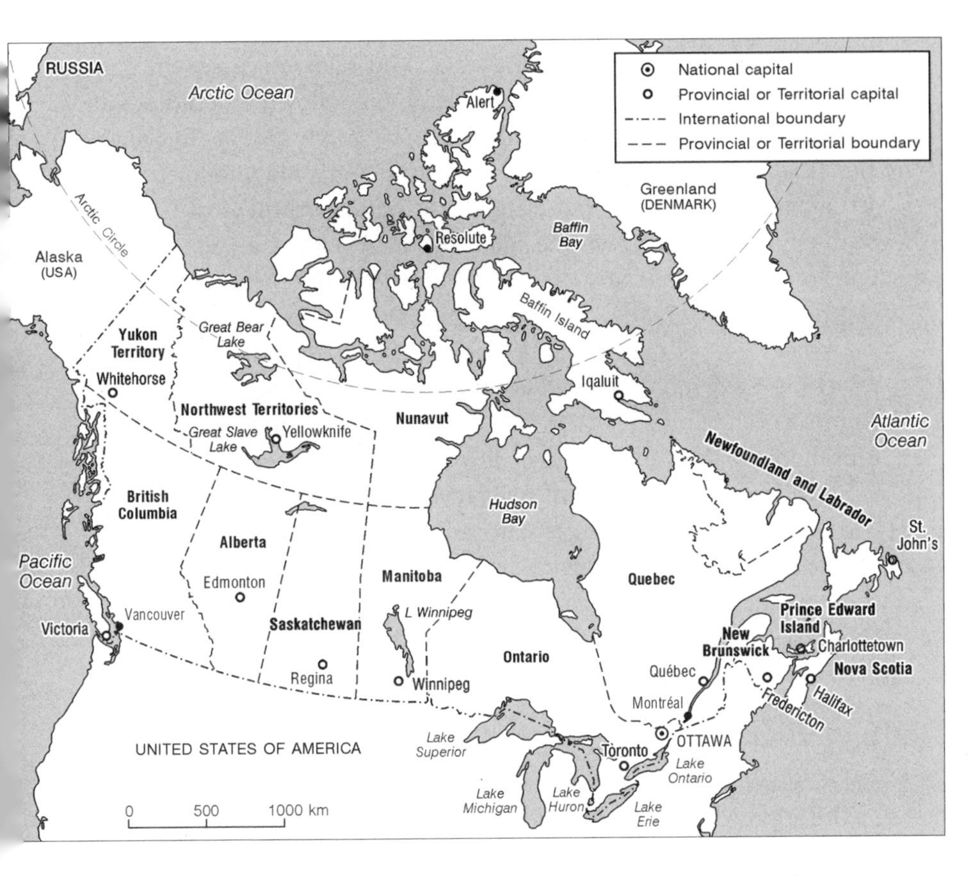

0.2 | Places mentioned in text

scientific practice, and a vision of the Canadian nation – that is, the Polar Continental Shelf Project (PCSP).[2]

The PCSP is a scientific organization, founded by the Government of Canada in 1958, that continues as a major operator in contemporary research in the High Arctic. Conceived as a Canadian nationalistic response to the International Geophysical Year, 1957–58, the PCSP held an initial mandate to render cartographically and geophysically the continental shelf to the north and west of the Canadian High Arctic Islands (figure 0.2). The PCSP was intended to reconfigure the nature of Arctic fieldwork through the maintenance of a long-term scientific project co-ordinating various governmental agencies. At the same time, the PCSP was to demonstrate Canadian sovereignty in the Arctic through maintaining *scientific* presence. This book discusses the consequences of achieving this vision during the following decades, and its contemporary instantiation through the PCSP base at Resolute.[3] The creation of the Territory of Nunavut in April 1999, following the settlement of the Inuit land claim with the Government of Canada, resulted in a major shift in the *political culture* of the Canadian Arctic. But how, if at all, has this affected the practices of environmental science in the region?

Debates across cultural anthropology, science studies, and geography have recently coalesced around the politics of science, the constitution of the human, and the construction of environmental futures. This book makes an empirical intervention into these discussions by examining the cultures of field science surrounding those tasked with understanding Arctic environments. It has become commonplace across the social sciences to state that globalization has recast socio-cultural practices and that this, in turn, requires new modes of analysis. This may well be the case, but changes in the globalized world have also radically recast the practices of environmental science. But what do the social worlds of the globalized science of environmental change look like? Such questions have rarely been subjected to empirical scrutiny. This is the task of this study.

The approach that I adopt here is along the lines of what Paul Rabinow has termed "an anthropology of the contemporary" (2008, 3). Through his studies of the modern biosciences, Rabinow has shown how, in such fields, emergent objects, capacities, and assemblages come to dominate practice (1996, 2003, 2008). In this book, I take further account of how environmental science *takes place* in the Arctic. The practices of the environmental and geophysical sciences have not enjoyed the same levels of research attention from social scientists as other arenas of modern science. To understand the contemporary field sciences, it is important to investigate their practices ethnographically. This requires "sustained research, patience, and new concepts." But it also requires, what Rabinow cryptically refers to as "modified old ones" (2008, 3). Examining contemporary practices in the geosciences suggests particular kinds of invention and the redeployment of ethnographic skills, capacities, and technologies. Analyzing the complexity of modern field science demands specific forms of social analysis, in which ethnography must remain central. These forms of ethnography, however, must be twisted and intertwined with other research competences for them to maintain continuing relevance and explanatory power.

In grasping my use of ethnography, I should state that I speak here from a contradictory position. This book forms part of a regional studies corpus – Northern and Arctic Studies – in which ethnographic expertise has been central to the development of all forms of social scientific research. At the same time, I come from a disciplinary tradition, geography, in which ethnography has been very much neglected since the field career of Franz Boas. Indeed, given that the contested relations between geography and anthropology can be traced to Boas's fieldwork in Cumberland Sound, I understand the precarious nature of connecting ethnography with spatial theories (Boas 1888; Powell 2015). In what follows, ethnography is understood not simply as a founding technique of a disciplinary community. Rather, I use ethnography to think through the spaces, networks, and practices that constitute Arctic environmental science

Much of the intellectual history of the Arctic has been about researchers collecting and depicting the lives of Inuit and other peoples. Indeed, sometimes, as in the case of Boas, these were scientists who became more interested in the cultural lives of the indigenous peoples of the Circumpolar Region than the landscapes they initially set out to study (Powell 2015). Although much has been written about nineteenth-century European explorers, and the associated mythical and romantic constructions of the Arctic, very little account has been taken of the scientists who spend so much of their lives in the North today. Moreover, the social histories of those other individuals, such as pilots, field assistants, and logistical staff, that are critical in the practice of environmental science have hitherto been completely hidden. To be clear, then, this book is articulating a broader notion of scientific practice that cannot be clearly demarcated between "science" and "logistics" in the field.

By concentrating on the different individual activities that compose science in the Arctic, the book argues for the importance of social analysis in understanding environmental research. The construction of Arctic science involves not only the enrolling of objects, but the development of contested forms of *subjectivity*. The Arctic scientific community, far from a homogenous grouping, is structured along power differentials in response to gender, class, and race. Furthermore, the human activities that create Arctic research in Canada are not limited to scientists, as conventional narratives would understand, but involve also a range of logistical personnel drawn from local Inuit, as well as migrant labour from Newfoundland and Quebec. In delineating this expanded conception of the *scientific human*, the book reveals the emotional dimensions of work and play in the Arctic. In doing so, the book advances arguments about conducting ethnographies of science that are informed by theories of practice and post-humanism.

Ethnographies of science, whether emerging from anthropology (Traweek 1996), sociology (Law 1994), or science studies (Latour and Woolgar 1986), began from studies of *laboratory* practices. There have been few ethnographic studies of field practices. Sections of this book therefore depict *what actually happens* during the quotidian production of field science in the Arctic. As the argument proceeds, though, I will demonstrate that such description is far from trivial and that the observation of field science has analytic, emotional, and ethical consequences. Moreover, I should stress that the book focuses on the emergent field cultures that are constructed by Arctic scientists in and through their interactions with other social actors. It is not enough to focus solely on particular traces, or moments during fieldwork, as in Latour's (1999) classic study of Amazonian pedologists. Rather, scientists and other social actors in the Arctic continually construct cultures that require serious examination.

In outlining the communities of practice behind global climate science, then, this study is classically ethnographic. There are cultural lives in the Arctic that are involved in some of the most urgent and demanding questions facing humankind. These lives have rarely, if ever, been subject to sociocultural analysis. In *Studying Arctic Fields*, I attempt to delineate this particular community of scientific practice.

Locating the Polar Continental Shelf Project

The scientific practices under study here have necessitated the incorporation of insights from both Geertzian and multi-site ethnography. Both the constitution of the Polar Continental Shelf Project as an institution and the consequent scientific practices enabled by it range across a number of places. As various participants in my study themselves pointed out, failure to comprehend this creates deficiencies in the understanding of Arctic environmental science. In this sense, my study draws on the tradition of Arctic ethnography that has emphasized continual travel and the making of connections *between* sites, which emerged originally from the study of the *seasonal* practices of nomadic Inuit (Pálsson 2001; Mauss 1906). I performed a multi-site institutional ethnography, focusing upon how the actors who constitute the PCSP are ranged across sites in the Arctic and southern Canada, and investigating the consequences of these distributions and experiences on the conduct of environmental science.

In the ethnographic present, the Polar Continental Shelf Project spans numerous sites, ranging from central administrative offices in Ottawa, major research bases in Resolute (Cornwallis Island) and Tuktoyaktuk (Mackenzie Delta), a building (which is infrequently staffed) at Eureka (Ellesmere Island), and co-ordinated scientific activities and networks of other dispersed buildings and field camps spanning the entire Canadian Arctic. Consequently, any attempt to study the PCSP must have a familiarity with these sites, and this, in turn, requires travel between the sites. At each place there are a number of individuals, each differently positioned, whose experiences and stories compose the contemporary activity of the PCSP. This would make extended dwelling at one particular site futile, whether at the research base at Resolute or in the offices at Booth Street, Ottawa. Moreover, the extent of research funding for science in the Canadian Arctic in recent years has been such that winter fieldwork has been impossible. Funds have not allowed PCSP research bases to be open year-round. This has prevented over-wintering by research teams, which has also affected the kinds of research question that can be investigated.[4] This also means that it would be impossible to study science in the Canadian Arctic through annual residence at any research base. It would, of course, be possible to spend the winter in the local, mainly Inuit community at Resolute Bay, but given the eight-kilometre distance between the settlement and the PCSP base, this would result in an entirely different understanding of Arctic scientific practice. The Inuit of Resolute have been studied by anthropologists (Tester and Kulchyski 1994; Marcus 1995), and there has been other excellent ethnographic work in other Nunavut communities (Stevenson 2014). The scientific community around Resolute has never been subjected to social investigation. My

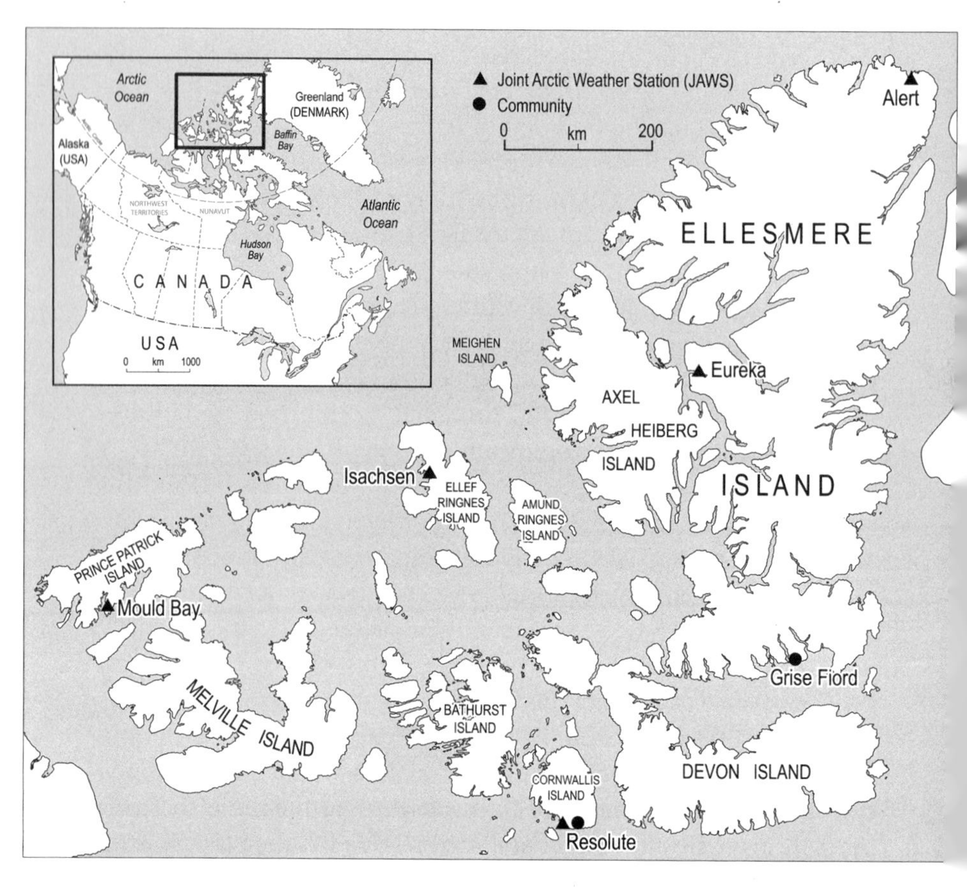

0.3 | Canadian Arctic Islands, indicating JAWS and Inuit communities

participant observation was therefore bifurcated over two substantial summer periods in 2001 and 2002, and focused on the scientific base and related networks.

Resolute has a remarkable position in Canadian *High* Arctic research, in that all researchers pass through Resolute airport and all Canadians, and the vast majority of non-Canadians, usually then stay at PCSP before flying out to the respective field sites (figure 0.3).[5] This meant that by spending long periods at PCSP Resolute I was able to come into contact with a huge range of researchers from, for example, Canadian universities, Danish research institutes, and even NASA. Moreover, Resolute is a staging area for almost all North American,

0.4 | Resolute Airport Terminal, July 2002

British, and Scandinavian expeditions to the North Pole, as well as for other forms of adventure tourism. The fact that Resolute is a place that is passed through as a means to an end (figure 0.4), rather than an end in itself,[6] meant that I could also observe the way in which many different social groups, including different bands of scientists, often did *not* converse with each other. For a relatively isolated community, therefore, there is a remarkable degree of human traffic, meaning that I could keep a low profile where necessary, such as when writing up daily field logs or conducting tape-recorded interviews.

My general activities around the base at Resolute included helping in the kitchen, talking to scientists, talking to employees, and generally "hanging out" in various parts of the camp. At other times, I conducted more formal interviews with scientists and staff at the base. As I will discuss, I spent significant amounts of time providing assistance in the field to various teams operating out of PCSP Resolute. Such teams would leave the base after breakfast and return in the evening for dinner, or occasionally, if the field site was close enough, return to the base for lunch, and leave again for an afternoon period in the field. On most days, scientists would also return to the field sites after dinner in the evening, and work until 10 p.m. or later, taking advantage of the optimal daylight conditions during the summer.

It was important, however, that I also visited other research sites across the High Arctic, lest I get a skewed impression of the nature of Arctic environmental science. The PCSP charters a number of fixed-wing, Twin Otter aircraft and helicopters to establish and supply field camps for scientific research across the Arctic. The field camps, located by the PCSP, are referred to as "fly-camps." Scientists would often have a number of scientific field sites in a particular area, and the fly-camp would be located in an optimal position as close to as many sites as possible, whilst taking into account the quality of landing access for aircraft, and other factors, such as fresh-water supply, natural shelter, and good visibility (for optimal polar bear detection). Consequently, I accompanied parties being taken out to their fly-camps or being "pulled out" and brought back to Resolute. On these trips I would help unload or load planes and perform general labour at the camp. On other occasions, I acted as a scientific field assistant in teams who were flying out to a field site and then returning later that day or night. These were "piggy-back" flights, in that a project of low priority would be taking advantage of a planned flight that was not carrying a full load. In these instances, I would collect samples or take measurements in stressful circumstances. On one occasion, for example, after a 3.5-hour flight to Mould Bay, Prince Patrick Island, we had around thirty minutes on the ground, whilst a Twin Otter was being loaded, to perform such tasks hastily, before boarding for the return flight.

Reflexivity in the Field: "Studying Up" and "Studying Down"

The PCSP involves a complicated network of different sites where environmental science is enacted. By choosing to investigate the actors whose practices together comprise the institution of the Polar Continental Shelf Project, I often encountered the difficulties that anthropologists refer to as consequences of "studying up" (Nader 1969; Ortner 2010). By this, I mean that the majority of the individuals whom I observed and interviewed were in more powerful positions than I was, and would have had the ability to hamper the completion of my project if they so wished.

In the first place, the Polar Continental Shelf Project is a branch of Natural Resources Canada, and thus of the Government of Canada. This presented all sorts of difficulties in getting permissions and access to bases and offices, especially as I am neither a Canadian citizen nor a landed resident of Canada. I should note that analogous obstacles occur in any attempt to write ethnographies of the institutions of government (Gusterson 1996). This aspect of studying up has meant that, although I have usually allowed the ethnographic narrative to flow, when reporting particularly contentious issues, I have referred to the documenting of the respective observation and conversation in my field notebooks.

Moreover, members of the Federal Science and Technology Committee were eager to know my recommendations for PCSP, especially as many scientists were lobbying for greater funding for northern science in general and PCSP in particular.[7] Similarly, given the complicated picture of northern science in Canada, with various non-governmental agencies and government departments involved in supporting and funding scientific activities, representatives from numerous organizations were interested in my research. An interview with one individual might involve numerous agendas, as the same person would often be involved in many of these organizations. Many retired scientists and employees of PCSP were also often now involved in other organizations, and thus had different objectives to push and wanted yet other stories to be told.

In the second place, the senior members of the PCSP hierarchy were interested both in why I wanted to conduct this research and in what I might find. I was asked after my second field period when conducting an interview with the PCSP director, for example, to provide guidance about regulations to be published by PCSP regarding their responsibilities and the duties of their support. Similarly, PCSP base managers, having been involved with the selection committees that made logistical decisions, knew of all the discussions regarding whether my project would be allowed to proceed. I was thus asked on a couple of occasions to state and defend my "hypothesis" at the most unexpected moments in the field.

In the third place, at Resolute itself, many PCSP-supported scientists wanted to know why I was there, not only because they were inquisitive about my project, but also because they wanted to know if I was being charged at "cost-recovery" rates because I was British. I will discuss the full implications of the *rules* of the PCSP in chapters 3 and 4, but this is an important marker of social status at Resolute. The limited resources available for northern science in Canada often meant that some people were initially cynical about my presence, until they learned who was footing the bill, upon which time they became much more interested. In this way, because of the manner in which I had applied successfully for PCSP support, my relationship to other scientists was rather different from what was common in most ethnographies of science. It was not obvious initially that I was not an environmental scientist, and the constant flow of scientists to and from fly-camps, and in and out of the field at Resolute meant that some individuals, perhaps on base for less than twenty-four hours, may never have known that I was writing an ethnography, because I did not have the opportunity to speak with them directly.

As most scientists are intelligent and highly educated, they were easily able to foresee potential consequences of this research project. Simon, a Government of Canada field scientist, articulated a commonly expressed view: "A history of Polar Shelf will be a good thing, because it could be used by people to help Polar Shelf survive future cuts." As Hugh Brody (1975) showed four decades ago in his ethnography of Inuit–white relationships in Iqaluit, the entry of the social scientist into "white" communities in the North has often been associated with ambivalence, because it is difficult to pigeonhole the researcher into a particular position. As Brody puts it, most "Whites are conscious that a social scientist spends his time uncovering information to which northern Whites usually have little or no access, a fact worsened by the confidentiality of that information, which can act only as an irritant to Whites who love gossip and are already nervous at having such an investigator in their midst" (72). Although Brody makes this argument from the perspective of the social scientist studying Inuit society, my relationship with scientists was often very similar, as I often discovered important, but confidential information about, for example, the future of PCSP funding.

Many of the interviews that I conducted were full of advice along the lines of how, "if I was writing on the history of PCSP, I would discuss this point or ignore that." Moreover, many of the retired field scientists could be subject to the usual problems of recollection, bias, and memory. However, there are still virtues in attempting to uncover such hidden spaces and voices from the field. One benefit of this sort of studying up is that the research subjects themselves are often fully aware of what researchers might find in the documentary record, and what they will not. For this reason alone, it is important to combine different research methodologies.

Moreover, scientists would provide a continuous commentary on my findings during my fieldwork. Over lunch, one scientist light-heartedly asked whether I had found any good scandals that he could use to get more support for the next season. Another asked about my methodology, and how a historian arbitrates between the stories of retired scientists and documents from the official records. A further suggestion was the submission of a questionnaire to the northern science community that stated explicitly that it came from *outside* the government. Others would stress to me that I should get this perspective or that. Most strangely, during a breakfast discussion about the vivid dreams some people were having, I was told that the second chapter of my book should discuss dreams. It was a common discussion topic at the base that the Magnetic North Pole somehow disrupted an individual's routine patterns of dreaming.

This interaction with scientists could also provide humour during meals on the base. Alex Jones, a young Canadian researcher, amused us all during one lunch with claims that "he's a spy" and by putting his hands over the ears of the "spy" whenever some trivial point, like what was on the menu for dinner, was mentioned. During this same meal, Cheryl Warner, another eastern Canadian, noted, "I am not sure about this guy, he just sits there and laughs." Steve Herbert (2001), in his study of the Los Angeles Police Department, makes a similar point about the experience of the young ethnographer in positions of unequal power relations.

More importantly in many ways, there were two instances in which I was, at the same time, "studying down" with those in less powerful positions in the field. As I will demonstrate, the seasonal employees of PCSP who work at the base are a vital set of voices for the history of PCSP. Although initially very wary of my presence because I was associated with the management, these employees began to see me as a way in which they could relay their anecdotes of the problems of the organization and respective individuals within it. In short, they began to see my project as a way in which to reaffirm the importance of the pursuits of their own lives by allowing the collection of their memories.

With the inhabitants of the hamlet of Resolute Bay, I was also studying down. Some members of the Inuit community are employed on seasonal contracts at the base, whether on cleaning or general labourer duties, and these individuals, as well as members of the hamlet council, were interested in what I had to say about the future presence of PCSP in the area, as it was an important part of the local economy.

In the chapters that follow, therefore, I was simultaneously studying up and studying down with different individuals during my fieldwork at the different sites that constitute PCSP in the High Arctic.

Studying Field Practices

Clifford Geertz famously noted that to understand any community it is necessary to study precisely "what the practitioners of it do" (1973, 5). In order that my discussion of the sociality of Arctic science be taken seriously, therefore, I demonstrate it through engagements with its community of practitioners. As such, I agree with Steven Shapin that such questions are best examined through the "historian's finest microscope," rather than the "social theorist's impenetrably airy generalities" (Shapin 1994, xix). At the same time, investigating practice requires engagement with theories of embodied action.

As historians of science have compellingly shown, the very point of modern laboratories is to demonstrate their supposed *placelessness* (Gieryn 2008; Kohler 2008; Gooday 2008; Withers 2009). This is nearby impossible for scientific *fieldwork*. Many students of scientific practice have failed to address this issue seriously (Powell 2007a). As a result, scholars have historically tended to neglect the peculiar difficulties of the field sciences. More recently, a vigorous historiography has developed concerned with redressing this lack of attention (Fischedick 1995; Hevly 1996; Kuklick and Kohler 1996; Outram 1996,

1999; Latour 1999; Bravo 2002; Kuklick 2011). As Dorinda Outram (1996, 1999) argues, fieldwork in the natural history tradition carried epistemological presumptions different from those of the cabinet and anatomy room precisely because it involved different conceptions of spatial experience (i.e., visceral movement *through* space as opposed to ascetic study *within* a bounded space). Many arguments about fieldwork have thus concentrated on the classic difficulty of any claims regarding travel – how do travellers, by their very nature involved in activities that lack credible witnesses, secure the *reliability* of any account? In consequence, travellers usually attempt to persuade by overloading accounts with specificity and abundant detail. This is, though, completely at odds with the ideals of the scientific revolution, because authority remains stubbornly reliant on the individual (Outram 1999). As such, many accounts of Enlightenment exploration had to rely on the authorial economy of physical endurance, courage, and bodily scars, or what geographer Michael Heffernan terms the "stigmata of truth" (2001, 216). For Henrika Kuklick and Robert Kohler, this reliance continues to reinforce the "ambiguous identity" of the modern field scientist by exacerbating a slippage between "amateur" and "professional" (1996, 11).

Another common theme in discussions of fieldwork is the epistemological appropriation of knowledges from field inhabitants by scientists (Kuklick and Kohler 1996). However, the quotidian practices involved in establishing commensurability between field scientists and indigenous residents are far from sociologically trivial (Bravo 1999). Moreover, actual instances of such encounters, precisely because of the methodological difficulties, have rarely been documented (Barnett 1998). As I will demonstrate, in order to recover the everyday processes of cross-cultural encounter in the field, it is necessary to deploy an ethnographic sensitivity. In doing so, some of the impacts upon the social worlds of both sides of cultural encounter begin to become apparent.

In his studies of field biology, historian of science Robert Kohler argues that the space of interaction, or "border zone," between laboratories and field spaces "is one of the most important in the cultural geography of modern science" (2002a, 1). The difficulties of reliable witnessing, replication, and cross-cultural encounter continue to result in the perceived lack of credibility of the modern field sciences relative to their laboratory-based counterparts. To deal with these issues, Kohler argues that practices are adapted by field scientists for specific places. Field biologists become skilled in recognizing how to use "place" to their advantage, such that the environmental variability of the field provides both the source of credibility problems for field biology and the resources for its solution. I will go on to discuss this further when considering the activities of field scientists in the Arctic, but it is important to state at the outset that deciphering such places is critical to understanding the practices of the field sciences.

Ethnographies of Practices

The approach adopted in this study conceives of scientific activity as composed of a number of interconnected practices. Recent work in human geography has often become preoccupied with a desire to study practices and performances (Thrift 2008). Scholars enervated by the excessive textualism of the cultural geography of the 1990s, manifest in studies of discourse and the politics of representation, have thus begun to reinstate the importance of materiality and become concerned with what have generally been termed non-representational "theories" or styles. However, as this book takes a different trajectory to the study of *practice*, some explanatory comments are required.

In this book, I wish to demonstrate, through ethnographic discussion, that the description of what actually happens has been misconstrued by such social scientists. Recording practices is crucial in understanding the role of situated social action. However, theorists of practice have themselves developed through an ethnographic sensitivity, which has in turn emerged from the study of quotidian activities of individuals in different social settings. That is, *theories of practice must emerge from observing actual practices.* As Latour (1999) argues, this dependence on the empirical origins of theoretical arguments is also "what science studies does best, that is, *paying close attention to the details of scientific practice*" (24; my emphases).

In what follows, I draw upon the notion of *communities of practice* (Lave and Wenger 1991). Drawing together arguments from cognitive psychology and educational theory, Jean Lave and Etienne Wenger develop a socialized notion of learning as situated activity. Individuals learn how to participate in groups of practitioners, or communities of practice, through *"legitimate peripheral participation"* until they are able to become fully competent members of the community (29; original emphases). Similarly, the ethnographer's task involves learning to participate in communities. Only then can the practices under investigation begin to be apprehended. As sociologist of science Harry Collins argues, "My own (partial) mastering [of] the practices of the communities I study is a very concrete achievement akin to what the members of those communities themselves achieve as they become members … [M]astery of a practice cannot be gained from books or other inanimate sources, but can *sometimes*, though not always, be gained by prolonged social interaction with members of the culture that embeds the practice" (2001, 107; my emphasis).

The purpose of an ethnographic account is to make more intelligible to others the actions of individuals within some sort of community. As Geertz (1973, 16) puts it, "The claim to attention of an ethnographic account does not rest on its author's ability to capture primitive facts in faraway places and carry them home like a mask or a carving, but on the degree to which he is able to clarify what goes on in such places, to reduce the puzzlement – what manner of men are these? – to which unfamiliar acts emerging out of unknown backgrounds naturally give rise."

It is through an ethnographer attempting to participate in the Arctic scientific community that the importance of a practice approach is made apparent. When the description of actual practices is undertaken, it is tempting to lapse into the recounting of tedious lists of apparently irrelevant incidents. However, through observations of quotidian activities in the Arctic, it quickly becomes evident that any notion of "scientific practice" cannot simply entail, say, the calibration of instruments or the measuring of geophysical phenomena, but must also centrally include the development of practical understandings and the structuring of social interactions between individuals. Such ethnography reveals important practices, such as work and play, that become central to Arctic science. As I discuss, it is precisely the human dimensions of social life, such as commitment, anger, and frustration, that most characterize the performance of scientific practices in the Arctic. It is for these reasons that I argue against post-humanist theories of practice, such as actor-network theory, that reduce the importance of the human subject in the development of field science. In composing this book, I am thereby attempting to show something of what the conduct of environmental science *means* to those who perform it.

At the same time, it is important to remember that if we wish to capture the importance of quotidian practices in contemporary societies, such as those of Arctic scientists, we need to develop ways in which the researcher can actually observe practices. It is not possible, whether in this case or, indeed, if it ever was, simply to *dwell* in a remote "field." The researcher must develop ways in which to render to other audiences the daily behaviour of the Arctic science community. The book therefore involves the deployment of methodologies in particular circumstances and under specific conditions that always require re-envisaging the epistemic basis of ethnography.

By being modest in my intentions for this ethnography, I am thus eager to develop an example of those "rich, complex narratives about who particular people are, what they think they are doing, to what end, and within what frame of meaning" (Luhrmann 2001, 3). As Luhrmann (4) argues, all ethnographers understand Geertz's "sensible point that simply describing what you saw was [and is] pretty useful." This book aims to accomplish this for Arctic field science in Canada.

Re-visioning Ethnographies of Science

If conceptualized loosely, "ethnography" has always been present in the geographical tradition, from tales of travel and exploration, through Carl Sauer and the Berkeley School (Sauer 1956), David Ley's (1974) deployment of participant observation in urban geography, to feminist research methodologies (Dyck 1993). However, Steve Herbert (2000) argues that geographers have generally misunderstood ethnography and that consequently its successful deployment has been rare in the discipline. For Herbert, ethnography is "a uniquely useful method for uncovering the *processes* and *meanings* that undergird sociospatial life" (550; original emphases). The researcher spends considerable time observing and interacting with a social group, and thus must both occupy the perspective of the group under study and maintain an analytical separateness as a social scientist. By focusing on what subjects *do* as well as what they *say*, ethnography maintains an analytic edge over methodologies that rely overly upon interviewing.

Ethnography thus allows both structure and agency to be studied in their instantiation in social practices, and as such facilitates the discovery of how social life becomes meaningful to those who enact it. However, this model of the need for anthropological fieldwork to result in "an ethnography" under relatively conservative strictures has come under increasing criticism over the past two or three decades (Marcus 1986). Not least, anthropologists have begun to see extensive dwelling in delimited sites to study particular "peoples" as increasingly impossible within the circulation of cultures under modernity.

The anthropologist George Marcus (1998, 2000) has outlined a way forward for ethnography under these conditions. For Marcus, ethnography needs to deploy a *multi-sited research imaginary*. Ethnography is viewed "as a research *imaginary* (provocations to alter or experiment with the orientations that govern existing practices) rather than a set of *methods* that are very specifically prescriptive for the conduct of field-work and writing" (Marcus 1998, 6; original emphases). These "multi-sited strategies" might include fieldwork in a number of different geographical sites or locations, tracing the connections between these sites (6). However, as an imaginary, Marcus is more concerned with how ethnographies are *conceptualized* rather than just the respective geographical locations. This conceptual work is difficult for Marcus, because the problematics of anthropological fieldwork have traditionally been pre-given. This resulted in a disciplinary body of work in which ethnography was "designed to be only description, or description as a form of argumentation, within the well-regulated discourse regimes of culture area" (12).

Much contemporary ethnographic research in anthropology or allied disciplines has been overwhelmed by a theoreticism that has been common across the humanities and social sciences over the past four decades. Such theoretical suppositions about cultural practices are, as I have discussed, almost wholly non-ethnographically derived. As Marcus puts it, the "anthropologist really does have to find something out she doesn't already know, and she has to do it in terms that ethnography permits in its own developed form of empiricism" (1998, 18). "Thick description" is therefore still critical in grasping the actualities of social practice. But when combined with a multi-sited research imaginary, it allows the reinvention of ethnographic practice, whilst at the same time protecting against over-theorization, which closes down the ability to capture the importance of actions by bodies-in-space. This research imaginary therefore "emphasizes the empirical challenge of just figuring out, demonstrating through description, and thus arguing for particular relationships and connections not at all obvious to the naturalized nominal categories of social space … or the theoretical stimulations which might have initially inspired an idea for ethnographic inquiry" (19).

For our purposes, then, a multi-sited imaginary in studies of field science provides a way ethnographically to contextualize the complex spaces of fieldwork, whilst keeping faith with the ability to capture information within the ethnographic narrative. In short, it provides us with renewed confidence in our quest to study scientific practices by allowing us freedom in ethnographic conceptualization.

However, it is important to remain open to the classical requirements of *detachment* and *reflexivity* during fieldwork. Although common problematics for the ethnographer, they become particularly acute in research on scientific laboratory practices (Latour and Woolgar 1986; Nader 1996; Traweek 1996). The ethnographer of science must work particularly hard at "the problem of maintaining analytic distance" (Latour and Woolgar 1986, 275). Such detachment, as Geertz (2000, 39) puts it, "is neither a natural gift nor a manufactured talent. It is a partial achievement laboriously earned and precariously maintained." For Latour and Woolgar, this detachment is difficult for ethnographers of science, because of the cultural baggage of "science" within their home communities. As John Law points out, in his fieldwork at a laboratory in northwest England, the proper consideration of reflexivity has ethnographic consequences: "Let me put it this way: as I describe the Laboratory *I do not always want to make myself invisible*. Thus I could offer an impersonal description of events in the Laboratory. I could talk of ethnographic research methods as if they were clear-cut, fixed and impersonal. I could pretend that there was no interaction between what I observed and myself as observer. But, as I've indicated, I believe that this would be wrong because ethnography is also a story of research – and in some measure a tale about the conduct of the ethnographer as well" (Law 1994, 4; original emphases).

Over the past two decades, a number of ethnographies of science have begun pushing forward the boundaries of disciplinary practice between anthropology and science studies (Gusterson 1996; Hayden 2003; Parry 2004; Green 2005; Masco 2006; de la Cadena and Lien 2015). I want to take insights from the notion of a multi-sited research imaginary and stress the importance of reflexivity, whilst guarding against the production of a "vanity" ethnography that is too much about my difficulties in conducting participant observation.

Moreover, any attempt to describe the everyday practices of other individuals inevitably entails the construction of a social relationship with the researcher. Describing the daily activities of scientists and PCSP staff is all very well, but I bear ethical responsibility for their representation in this book. At the same time, precisely because of the ethnographic method deployed, it is necessary to be *reflexive* about these descriptions. It is a commonplace of ethnography that the presence of the outsider-as-researcher influences the practices that are observed. My presence in the field, as passages later in the book might suggest, allowed individuals to air all sorts of views that might never have been articulated had I not been writing about the field practices of PCSP. It will certainly appear in what follows that people may have wanted to speak to me as a conduit for the expression of frustrations about the current state of PCSP and northern science in Canada. There have been difficult ethical issues to negotiate, therefore, as I have tried to remain faithful to these competing voices, whilst not revealing their individual origins, and allowing my own interpretations to be recorded.

Structure of the Book

The book is structured as follows. Chapter 1 investigates the founding of the Polar Continental Shelf Project amid debates about Canadian nordicity and national identity. As such, the chapter discusses the PCSP as a mission of *scientific sovereignty* in the Canadian Arctic. Prime Minister of Canada John Diefenbaker's unprecedented electoral success in March 1958, based on his "Northern Vision," led to demands for geophysical and cartographic activities to demonstrate Canadian knowledge and presence over the areas northwest of the High Arctic Islands. The PCSP was envisaged as part of a wider geographical project to bring the High Arctic into the Canadian national imagination. This vision of attaining sovereignty through scientific activity is developed through its connection to the role of northern nationalisms in Canada. It is argued that this particular constellation of scientific nationhood created a persistent *culture of temperance* within the PCSP.

Chapter 2 discusses how the PCSP was construed by its field scientists, at the same time, as an attempt to reorganize and reconceptualize scientific research in the Arctic. The introduction of experimental techniques into Arctic fieldwork is discussed, as scientists developed notions of "structured observational measurements" and "field experiments." These debates are situated in the epistemic context of Cold War philosophies of field science. But, as I show, such practices also always interacted with difficult environments and there were emotional consequences to these physical impediments.

Chapter 3 introduces material derived from participant observation of the practices that constitute the contemporary PCSP by outlining the organizational structure, institutional rules, and spatial layout of the base at Resolute. It examines how this spatiality reflects the origins of PCSP, as discussed in chapter 1, and how it is maintained by the PCSP management through their rules for conduct in the field. It is argued that the spatialities of the base in Resolute help constitute contemporary cultures of temperance within the PCSP. Notions of proper field practice and Canadian nationalism are reformulated through the contemporary activities of the PCSP.

Most crucial in the conduct of scientific fieldwork at PCSP Resolute is the performance of an identity of the *good field person*. Chapter 4 discusses the importance of such constructions of scientific selves in the Arctic by drawing from participant observation with scientific field teams. Drawing further from conversations and interviews with scientists, this chapter reveals how Arctic scientists themselves establish a certain subjectivity that involves engagement with policy and management. This appears to rejuvenate the tradition of the holistic Arctic investigator within contemporary environmental science.

The fifth chapter discusses social interactions between scientists and Inuit in the field. Resolute, currently the logistical centre of scientific research in the Arctic, is also the site of a relocated Inuit community, moved by the Canadian state during the early 1950s. Scientists, and other actors involved in the constitution of science, thus become signifying traces of the state. However, for reasons of logistical expediency and colonial legacy, these two communities rarely come into contact.

These arrangements change for a single day each year, Canada Day, when Canadians are encouraged to celebrate their nationhood. All too often, anthropological understandings of Arctic cultures have instantiated and memorialized particular indigenous cultures. Similarly, social studies of science have rarely understood the ethnographic specificity of cultural encounters during instances of field practice. The chapter depicts the social spaces of the Arctic during the carnivalesque occurrences of Canada Day. By drawing Victor Turner's notion of social dramas into conversation with Sherry Ortner's practice theories, this chapter stresses the importance of *encounter* between different groups of socio-cultural actors in the enactment of modern environmental science.

Chapter 6 investigates the consequences of PCSP's contemporary role as a logistics organization for Arctic science. By deploying the notion of Arctic play, a concept that involves reworking some of the classic tropes of Arctic sociology, this chapter examines the emotional repercussions of facilitating science for employees of the PCSP. It shows how logistical staff, through their quotidian labours, constitute themselves as the "real scientists" in the Arctic. In doing so, this chapter complicates prevalent conceptions of the Arctic scientist. In the process, a number of theoretical observations are made about the business of observing scientific practices.

The cultures of labour involved at other levels of the PCSP employment structures, as well as accompanying cultures of science, are discussed in chapter 7. The base managers, through their rules that constitute Resolute's culture of temperance, are shown to have a critical role in the conduct of environmental science in the Arctic. In performing these practices of scientific governance, the base managers perpetuate an earlier vision of scientific sovereignty. Frameworks that govern science in the Canadian Arctic, such as the gendering of field science and Inuit participation in fieldwork, are also investigated.

The book concludes with an epilogue that brings the narrative into the future, as well as summarizing the argument. Analyzing the conduct of scientific practice in the Arctic by granting attention to the individuals involved in its conduct and performance reveals much about understandings of the scientific human. By focusing on what actually happens during the everyday practices of the Arctic scientific community, the emotional dimensions of work and play become evident. It is only through these dimensions that a fuller understanding of field science is possible. The book thereby valorizes the continuing need for ethnographic sensitivity in studies of scientific practice.

1 Scientific Sovereignty, Nordicity, and the Canadian Nation

We ask from you a mandate; a new and a stronger mandate, to pursue the planning and to carry to fruition our new national development programme for Canada … This national development policy will create a new sense of national purpose and national destiny. One Canada. One Canada, wherein Canadians will have preserved to them the control of their own economic and political destiny. Sir John A. Macdonald gave his life to this party. He opened the West. He saw Canada from East to West. *I see a new Canada – a Canada of the North* … There is a new imagination now. The Arctic. We intend to carry out the legislative programme of Arctic research, to develop Arctic routes, to develop those last hidden resources the last few years have revealed … *To the young men and women of this nation I say, Canada is within your hands. Adventure. Adventure to the nation's utmost bounds, to strive, to seek, to find, and not to yield. The policies that will be placed before the people of Canada in this campaign will be ones that will ensure that today and this century will belong to Canada. The destination is one Canada.*

Prime Minister John Diefenbaker, 12 February 1958, "A New Vision"

Introduction

Mapping territory allows for the completion of the epistemic project of the nation. But in order to demonstrate domination over territory, sovereignty must become accepted by global audiences. In Canada, uncertainties of delineation have resulted in a succession of cartographic projects in the Arctic. In the twentieth century, this process began to occur through the establishment of *scientific sovereignty*. Histories of scientific sovereignty are evident across different regions of the world. For the Canadian Arctic, the Polar Continental Shelf Project forms a major epoch in this history. Moreover, Canadian nationalism has often been reliant upon an imaginative geography forged upon the vision of the northern landscapes of the country. Canadian intellectuals have understood this imaginary as *nordicité canadienne* (*Canadian nordicity*). This is an imagination that has been critical in attempts to create a *Canadian* nation, and it was crucial in the founding vision of the Polar Continental Shelf Project. Indeed, the scientific practices of the PCSP have been completely intertwined with the historical geographies of a national project in Canada.

In this chapter, I will discuss the founding of the Polar Continental Shelf Project within the context of these wider debates about Canadian nationhood. As Benedict Anderson (1991, 6) puts it, a nation "is an imagined political community – and imagined as both inherently limited and sovereign." Edward Said (1978) has shown how geographical imaginaries operate in cultural constructions of peoples and environments. In order to understand how this imaginative geography influenced the PCSP, I will first discuss how nordicity emerged as an unifying construct during nineteenth-century debates over the Canadian national imagination. The traditional practices of the field sciences were critical in attempts to develop *national* communities. In the Arctic those activities, such as "collecting, sketching, measuring, recording, [and] classifying," remained important throughout the twentieth century (Bravo and Sörlin 2002, 18). Indeed, in the Canadian Arctic, it was practices undertaken by PCSP scientists, such as mapping and surveying, that became critical in attempts to develop a postwar *pan-Canadian* national identity.

Even more importantly, these histories indicate a time when to be a *good field scientist* entailed the same goals as being a *better Canadian*.

As will be pursued in later chapters, the association of a particular form of Canadian nationalism, public service, and scientific practice continues to have legacies in the present activities of the PCSP.

Although the Arctic was of less immediate concern to Canadians during the turbulent middle decades of the last century, by the late 1950s events had brought the region back to centre stage. The International Geophysical Year, 1957–58, and the Law of the Sea Conference in Geneva, 1958, had changed the scientific, legal, and political context of the Canadian Arctic. Although the International Geophysical Year was initially conceived by some as an initiative for scientific co-operation and global peace, there were also nationalistic undertones in countries such as Canada. As nordicity reached its political and popular apogee with the federal electoral success of John Diefenbaker in 1958, field scientists were sent to the High Arctic literally to expand the bounds of Canada. The establishment of the PCSP involved strain between notions of global science and those of national imagination. This chapter illustrates the tensions between an idealized scientific globalism, strategic continentalism, and territorial sovereignty.

These tensions are recovered from oral historical interviews with field scientists involved in High Arctic fieldwork in the late 1950s and early 1960s, who participated in the Polar Continental Shelf Project. By drawing *both* from interviews with the fieldworkers and policy-makers involved, *and* the conventional archival record, I am able to show the potentialities that are opened up for histories of the field sciences in using oral history, as well as recovering some of the vivacity of field practice.

Constructing Northern Nations: Imagining Canada

It has become something of a commonplace to state that histories of the polar regions have usually been written by scholars attached to specialist institutes, such as the Scott Polar Research Institute or the Arctic Institute of North America, University of Calgary (Spufford 1996; David 2000). This has had a double-sided consequence. On the one hand, polar historiography has generally tended towards hagiography, with accounts concentrating on chronological narratives of the exploits of explorers with little attempt to link into wider theoretical and substantive debates. And, on the other, it has resulted in the neglect of the polar regions by academic historians, so that accounts of, say, nineteenth-century colonialism have failed to consider sufficiently these areas of the world. This is obviously an over-simplification, and there have been a number of recent scholarly studies of polar histories (Jones 2003; Robinson 2006; Hill 2008). Nonetheless, there is still much work to be done in order that histories of polar science and exploration receive the attention they deserve. Moreover, accounts of the polar regions have hardly begun to indicate how scientific practices become intertwined with national projects.

In order to understand how geographies of northern landscapes were used to develop a national identity in Canada, it is necessary to grasp a broader notion – what Sverker Sörlin (2002, 89) terms "the idea of civilization's northward journey." This notion was strengthened through nineteenth-century anthropological and geographical debates about the supposed climatic determinants of racial characteristics, but Sörlin shows that it has older origins in arguments about the composition of the Swedish nation. In Sweden, the conception of a northern destiny was critical in nation-building initiatives and attracted many young individuals to the cataloguing of natural resources in the second half of the nineteenth century. These practices of collecting and recording were central to the field sciences. Although always present in the cultural imagination to some degree, this stress on the northern future of the Swedish nation emerged forcefully *whenever national identity was under a perceived threat*, such as at the end of the union with Norway in the early twentieth century (Sörlin 2002).

At the same time as defining national identities, such field practices were also *colonial* inscriptions of ownership in defiance of both indigenous inhabitants and the competing claims of other European states. As Kirsten Thisted argues, there are comparative debates about national unity in Denmark, although with starker colonial overtones because of the location of the "northern territory" of Greenland (2002). It is this complex relationship between national identity and imperial interest that has led Gísli Pálsson to develop the "notion of arcticality" which made "the Arctic both exotic and domestic" (Pálsson 2002, 275; 2001). Drawing an obvious parallel with Said's (1978; 1993) seminal work on "Orientalism," and David Arnold's (1996) sketching of the cultural construction of "tropicality," Pálsson argues that *arcticality* is evident across Scandinavian engagements with northern territories.

In what follows, I will show how such discourses continue to circulate through scientific practices in the Canadian Arctic.

Supporting Homi Bhabha's (1994) arguments regarding colonial cultures, then, Canadians are *ambivalent* about their northern environments. As a character remarks in Mordecai Richler's discussion of some of the founding myths of Canada in his monumental novel *Solomon Gursky Was Here*, Canadians have always had difficulty in sustaining any sort of imagined national community:

> Canada is not so much a country as a holding tank filled with the disgruntled progeny of defeated peoples. French-Canadians consumed by self-pity; the descendants of the Scots who fled the Duke of Cumberland; Irish the famine; and Jews the Black Hundreds. Then there are the peasants from the Ukraine, Poland, Italy and Greece, convenient to grow wheat and dig out the ore and swing the hammers and run the restaurants, but otherwise to be kept in their place. Most of us are still huddled tight to the border, looking into the candy store window, scared by the Americans on one side and the bush on the other. And now that we are here, prospering, we do our best to exclude more ill-bred newcomers, because they remind us of our own mean origins in the draper's shop in Inverness or the *shetl* or the bog. (1989, 398–9)

At the same time, since Confederation in 1867, many intellectuals have undertaken "the eternal Canadian search for identity" (Henighan 2002, 35). As Richler implies, this may be a partial consequence of the continual sustenance of the Canadian population through large-scale immigration. However, it is also crucially important that Canada is a "post-colonial nation" (Hulan 2002, 8). There has thus been a perceived need to sever ties from the imperial metropoles in Britain and France and develop an official unitary narrative. But as many scholars have argued, Canada still perpetuates many cultural formations typical of a colonial society. It is this "colonial mentality" that, for Margaret Atwood, has impeded the development of Canadian literary culture, with undue stress being placed on the literature of the United Kingdom, United States, and France (2003, 50). Stephen Henighan, in a collection of essays short-listed for the Governor General's Award for non-fiction in 2002, argues that Canadian thinkers still have difficulty in expressing cultural authority, and maintain "a latent desire to dispense with the awkwardness of being Canadian and meld into some larger, simpler entity" (2002, 135). When Atwood was invited to deliver the Clarendon Lectures in English Literature at Oxford in 1991, her introduction suggested this ambivalence: "Canada – lacking the exoticism of Africa, the strange fauna of Australia, or the romance of India – still tends to occupy the bottom rung on the status ladder of ex-British colonies" (1995, 2). It is within this context that Canadians have culturally constructed the North.

Nordicity and the Quest for National Unity

Attempts to construct a Canadian culture, then, have often relied on the imaginative circulation of the diversity of Canadian physical environments. Many intellectuals have been eager to define an imagined community for the nation, and in so doing almost all have seen *geography* as the register of Canadian distinctiveness. As such, the Canadian geographical imagination has encompassed a relatively high degree of what we might term "environmental determinism."[1] The political economist Harold Innis, in his magisterial economic history of the fur trade published in 1930, argued that the production of semi-processed raw materials for export, such as beaver pelts, led to the development of transportation and communications systems to access a huge resource hinterland for the economic core along the St Lawrence River (Innis [1930] 1970). For Innis, therefore, a transcontinental nation of Canada developed *because* of its geography. In 1936, Prime Minister William Lyon Mackenzie King famously mused, "If some countries have too much history, we have too much geography."

Most often, the crucial features of Canadian geography were taken to be its *northern* landscapes. In the half-century after Confederation, this use of northern geography for the purposes of nation-building was most associated with the Canada First movement (Berger 1966; Grace 2001; Hulan 2002). The agenda of this group of nationalists was outlined by Robert G. Haliburton in an 1869 lecture to the Montreal Literary Club: *The Men of the North and Their Place in History*. For Haliburton, the ethnic heritage from northern Europe, together with the severe winters and latitudinal position of the new Canadian nation, combined to form a *northern destiny*. For these nationalists, liberty itself and all its institutions found their origins in northern societies (Berger 1996). Canadians were to become the northern pioneers of the new world, and in the process would develop a national character based on *temperance* borne of strength, vigour, and purity (Berger 1966).

The distasteful masculinist and proto-racist undertones of the Canada First movement are obvious, and have been usefully discussed by Sherrill Grace (2001). However, although now less starkly drawn, this northern imagination continues to act as an epistemic justification for southern Canadian intrusion into indigenous territories. As Grace puts it, the relationship between southern Canada and the North is very much "a history of civil imperialism" (1996, 2). What I want to stress here, however, is that this myth of *nordicity* or *nordicité* attempted to provide the imaginative grounds for the fundamental unity of a Canadian nation that transcended what novelist Hugh MacLennan famously termed the "two solitudes" of French and English Canada (Berger 1966; MacLennan [1945] 1967). Moreover, as well as providing a unification theme for the New Dominion, and providing for *both* inheritance *and* separateness from the colonial motherlands, the northern ethos was also able to indicate that Canada was necessarily distinct from the United States. As critic Christy Collis (2003, 155) puts it rather neatly, "It is by now all but axiomatic that one of the core ideas that coheres the sprawling imagined community of the Canadian nation is nordicity: the geographically determinist notion that Canadians are unique as a national culture because of their relationship with the north."

As Grace suggests, the Canadian north becomes a "*discursive formation*" relying on "a plurality of ideas of North that are in constant flux yet are persistent over time, [and] across a very wide field of endeavour" (Grace 2001, xiii; original emphasis). This imaginative north circulates across a number of Canadian cultural registers. After the Great War of 1914–18, the depiction of northern landscapes became a focus for the Group of Seven impressionist artists, such as A.Y. Jackson and F.H. Varley. Similarly, Glenn Gould used the North to reinvent the medium of the radio documentary in the 1960s, through the representation of competing voices in his "Idea of North" program broadcast by the Canadian Broadcasting Corporation (Gould [1967] 1985).

Nordicity has played a huge role in Canadian literature because it contributes to "dreams of what the country once was or might become" (Coates and Morrison 1996, 18). Consequently, most of the country's intellectuals "have contributed their 'northern' book, perhaps as a Canadian rite of passage" (13). Sherrill Grace prefaces her book with the claim that it was written "because I am Canadian and my love for and desire to understand this stubborn, complex, infuriating place that I call home drives me" (2001, xi). For Alison Mitcham, "Perhaps the most exciting creative force in contemporary Canadian fiction – French and English – is the Northern Imagination" (1983, 9). This imagination was marshalled most effectively by Atwood in her attempts to demonstrate the rightful concerns of a *Canadian* literature, or "Canlit," in defiance of English and American literary imperialism (Atwood 1972, 17). "Canadian literature," Atwood argues, "as a whole tends to be, to the English literary mind, what Canadian geography itself used to be: an unexplored and uninteresting wasteland, punctuated by a few rocks, bogs, and stumps" (1995, 2).

Indeed, by the late 1970s, many practitioners of the discipline of geography in Canada became preoccupied with attempting to *measure* nordicity along the lines presented by Québécois geographer Louis-Edmond Hamelin (1979, 1988a, 1988b). For Hamelin, nordicity was a function of different aspects. An index was therefore to be calculated by attributing scores out of 100, defined as "*valeurs polaires* or VAPO," for ten components, such as latitude, various climatic factors, population density, and economic activity (1979, 18). The sum of the ten VAPO scores gives an index of nordicity out of 1,000. And thus, "The maximum of 1,000 VAPO is theoretically attained at the Pole; hence one VAPO represents one-thousandth of the polar maximum" (18). After calculating these values, Hamelin constructed maps with isolines of nordicity, termed "isonords" (27). Isonord maps allowed a broad four-way division of the Canadian territory into "Base Canada" along the US border, the "Middle North" of the provincial hinterlands, the "Far North" across the top of mainland Canada and including Baffin Island, and the "Extreme North" of the High Arctic Archipelago (xiv). Moreover, Hamelin argued that indices of nordicity vary over time, such as with the development of new communication routes or the opening of a mine. His purpose, therefore, was to *quantify* "the progressive denordification of Canada" through social, technological, and economic developments (281). As the North became quantifiably more hospitable to southern Canadians, then it would be incorporated further into the national imagination. As Hamelin noted, "Nordicity is more than simply a new term, it is an entire programme" (283).

During the 1980s and 1990s, a number of writers used the Canadian North to undertake voyages of personal and intellectual discovery. These accounts combined travel writing, creative writing, and literary criticism, and examples include Rudy Wiebe (1989), John Moss (1996) and Aritha van Herk (1990). For critic and endurance runner John Moss, the purpose of this project is the very understanding of the meaning of geography: "Trying to define geography: the imposition of knowledge on experience in a specified landscape … The struggle to define geography is a question of being. Where in the world are we? We yearn for a familiar perspective. We need geography, it seems. We invent it" (1996, 1–2).

Moreover, for Moss, *landscape* is "the antithesis of geography" (1996, 5). This definition of geography is rather problematic from a disciplinary perspective but results from the influence of Hamelin's project. Any attempt to impose knowledge over experience of the Arctic landscape is, under Moss's phenomenological conception, geography, and all those who travel to the Arctic thus *become* geographers. However, it is important that the geography of the Arctic, even for Moss, is a *national* project, because "Canada, unhyphenated, held possible in imagination, reaches to the North Pole itself" (136).

More remarkable in this vein of Arctic geographies written by literary critics are Aritha van Herk's feminist travel accounts. Van Herk is much more critical of the association of the Canadian North with national identity: "The Arctic … is a mirage of land and water, an ice-shape beyond political declension. The farthest reaches of the north are a configuration of the imagination, a transgressive act for those who dream of a nirvana beyond ice" (1997, 79). By playing with this relationship between geographical imagination and fiction in the Arctic, van Herk has attempted to develop a notion of "geografictione" (1990, 87). As the Arctic is an area for *transgression*, Ellesmere Island becomes a geografictione, which, for van Herk, is a place where blank spaces allow the rearticulation of narratives and, ultimately, for female emancipation. In her *Places Far from Ellesmere*, therefore, the island becomes both the escape for van Herk's transitory life across Alberta as well as for Tolstoy's Anna Karenina, because *Anna Karenina* is the only book she could carry when hiking on northern Ellesmere. Thus, for van Herk, "Anna can invent herself in an undocumented landscape, an undetermined fiction" (125).

These accounts are provocative and intentionally ironic, but I must note the apparent lack of respect for *existing* Inuit narratives of Ellesmere Island. Although often neglected, Inuit representations of their homeland have become increasingly common (Ipellie 1993, 1997; Petrone 1988, 1991), and a growing number of accounts by southern Canadians give serious attention to the narratives of northern indigenous peoples (Cruikshank 1990; 1998; 2005; Wachowich 1999). Reviewing these cultural accounts of the North provides an important context for the imaginative construction of the region within scientific-political agendas.

The Return of a Northern National Identity

After the early 1970s in Canada, moves towards indigenous self-determination following the Mackenzie Valley Pipeline Inquiry, as well as the introduction of official *multiculturalism* under the Liberal governments of Pierre Trudeau in the 1970s, meant that the North was less often deployed as a unifying myth. However, by the 1990s, with the signing of the North American Free Trade Agreement with the United States and Mexico and the second Quebec sovereignty referendum of 1995, the existence of the Canadian polity again came under perceived threat. This resulted in a re-energized debate over Canadian political futures (Saul 1997, 2003; Griffiths 2000; Kymlicka 2001, 2003).

The relatively successful adaptation of political and legal structures to ethnic and regional diversity has been seen by many commentators as a remarkably little-celebrated dimension of the "unmarketable obscurity of being Canadian" (Henighan 2002, 107). For political philosopher Will Kymlicka, identity politics characterizes contemporary Canadian multiculturalism so much so that it becomes banal (Kymlicka 2003). Canada, as Hulan puts it, could be "the first nation in which incredulity toward national identity actually comes to define and to legitimate the state" (2002, 187).

It is in this context, where the Canadian polity is under perceived threat but there is a firm commitment to political multiculturalism, that a reformulated argument for nordicity has resurfaced. This draws a Canadian identity together across French, English, and indigenous Canada, again based on human relationships with northern environments. John Ralston Saul, a leading Canadian "public intellectual" and consort of former governor-general of Canada Adrienne Clarkson, has recently used nordicity to argue for a distinctive *animism* at the heart of Canadian public culture, by which a *co-operation* between people and environment becomes distinct from the domination over "nature" more common in the United States (Saul 1997). For Saul, the physical environments of Canada have prevented the sustenance of any argument for manifest destiny. Drawing across arguments of early federalist Canadians, as well as the Inuit notion of *isuma* – that is, intelligence that comes from knowledge of the individual's social and environmental responsibilities – Saul argues that Canada is necessarily a "nuanced, complex and relatively decentralized society" (107). Atwood (1995) similarly tries to recover a northern environmental ethic for Canadians from indigenous peoples. Saul uses this notion to define Canadian political culture: "You could call it the model of the marginal or frontier or northern nation-state … Given the uncontrollable nature of the place, this was a model which was not intended to achieve the sort of static state implicit in the mythology of domination. This is a nation conceived as existing in permanent motion; more a sensibility than an ambition" (1997, 107).

For Saul, Canada is "the first of the post-modern nation-states," a polity emerging from the complexity of "the fundamental three-part foundation of the country: the Aboriginals, the Francophones, and the Anglophones" (2003, 24, 18). As John Moss argues, this revivified Canadian nationalism, based on complexity resulting from northern environments, may be "a Canadian nationalism that will make some readers uneasy. It is not, however, a political nationalism but rather a desire to examine and share the genius of Canadian experience from a particular perspective, Canada as a country rather than a state, as landscape rather than geography, as a culture of infinite particularities, a community of endless diversity, a lovely and necessary and breathtakingly beautiful land" (1999, vii).

Having shown how these arguments about nation and nordicity have been deployed across Canadian culture and circulated through conceptions of northern Canada, I now move to discuss their role in the foundation of the Polar Continental Shelf Project during the scientific and geopolitical situation of the 1950s.

Canada and the International Geophysical Year, 1957–1958

The International Geophysical Year (IGY), which ran between June 1957 and December 1958, involved participants from over sixty nations ranging across the earth and environmental sciences (Sullivan 1961). The Executive Committee of the UN World Meteorological Organization suggested that a widespread International Geophysical Year would be more productive than the initially proposed bipolar focus. Consequently, in 1952, the International Council of Scientific Unions announced the IGY. The IGY program resulted in observations conducted in thirteen scientific fields, including geomagnetism, ionospheric physics, and oceanography. Despite the broad range of disciplines involved, the activities always maintained their original connection with polar science.

The IGY resulted in scientific landmarks and unprecedented cooperation between the United States and USSR. The launch of *Sputnik 1* and *Sputnik 2* in October and November 1957, as part of the USSR's satellite program, greatly exacerbated Western fears about potential missile threats. *Sputnik* also undermined Western assumptions regarding scientific and technological superiority (Smith 2000). Ultimately, *Sputnik* resulted in geophysical scientists being elevated to greater standing within political circles, thus recasting the post-war settlement between science and politics (Doel and Harper 2006). This scientific focus was also to have productive results for international diplomatic relations. The success of the Antarctic geophysical program during IGY facilitated the resolution of long-standing sovereignty disputes in the signing of the Antarctic Treaty in 1959. But this linking of geophysics and sovereignty hints at a contradiction at the heart of the IGY – attempts at international scientific cooperation were always concurrent with Cold War national rivalries.

Moreover, such competitive tensions were not limited to the superpowers, and the IGY helped focus widespread attention towards scientific competition amongst the "middle powers," such as Canada. The Canadian research contribution to IGY involved programs at seventy-six stations. These research sites included twenty-six in the Canadian Arctic that resulted in major contributions towards understandings of atmospheric physics, glacial ice regimes, and magnetic interference in northern radio communications. Notable amongst the Canadian IGY initiatives was Operation Hazen, operated by the Defence Research Board on northern Ellesmere Island. The base at Lake Hazen provided for research in glaciology and meteorology, with supplementary studies in geology and archaeology. However, the consequences that the IGY had for the role of scientific practices within the Canadian political imaginary have been neglected by historians.

During the 1950s, Canada sought to distinguish itself from both the United States and the United Kingdom by adopting "a distinctive role as a 'middle power' in world affairs" (Eldridge 1997, xii). This occurred in the context of a post-colonial reconfiguration of the global order during the later years of the decade, as the United Nations began to play an important role in geopolitics. It is within initiatives like the IGY that it is possible to detect an embryonic scientific nationalism across some of these "middle powers." Canada was no exception.

In what follows, I examine these issues through an investigation of the establishment of scientific field practices within the broader context of Canadian political nationalism during the late 1950s. This is accomplished by exploring one of the few legacies of the IGY for Canadian *science* policy, namely, the founding of the PCSP. Expanding on the earlier discussion of nordicity, the next sections consider the attempts to mobilize a pan-Canadian nationalism in response to perceived American and Soviet incursions upon territorial sovereignty during the IGY. Once successfully harnessed, I will argue, this political nationalism was made manifest through specific forms of logistical and scientific practices in the Arctic. The PCSP's ostensible purpose was to collect geophysical data for US satellite launch tests in the Canadian High Arctic. However, by examining the geopolitical context of the UN Conference on the Law of the Sea in 1958, together with the unprecedented electoral success of the Progressive Conservatives across Canada during the same year, the story of the founding of this Arctic field science organization becomes of much wider significance.

John Diefenbaker, Nationalism, and Continentalism

For many scholars, the political and economic achievements of Canada during the Second World War helped develop a new sense of Canadian nationalism. After 1945, there was a widespread desire amongst both policy-makers and the public service to facilitate what diplomatic historian Greg Donaghy terms Canada's transformation from "colony to nation" (1998, 18). The early Cold War period of the late 1940s and 1950s, according to many students of Canadian politics, was the "so-called 'golden age'" of Canadian diplomacy (Campbell 1998, 12). Among the many perceived achievements of Canada in these years was the active pursuit of "multilateralism," mediation, and firm participation in the newly formed United Nations. During the Suez Crisis of 1956,[2] Canadian minister of external affairs Lester B. Pearson played a decisive role in the UN negotiations for peaceful withdrawal from Egypt, and was subsequently awarded the Nobel Peace Prize in 1957. Such activities increased Canada's standing in the world, as it came to be regarded as an "honest broker" on the international stage. Recent scholarship has shown that Canadians acted with more self-interest than was generally recognized, but the diplomacy of the early Cold War remains a powerful national narrative that is often contrasted favourably against US foreign policy.

However, for all the possibilities created for Canadian nationalism by the decline of Britain and France, the resurgence of North American continentalism provided a new challenge. Colonial dependence upon Britain was being replaced, by the late 1940s, by a neo-colonial relationship with the United States (Drache 1970; Resnick 1970). The tensions between forces for continentalism, and the interlinked fears of US supremacy, as against Canadian nationalism, have been well noted at different periods during the last century or so (Moffett 1907; Stuart 1994; Stairs 1998). The years of the early Cold War recast this relationship again.

Fears of American incursions upon Canadian sovereignty were exacerbated during discussions over the TransCanada pipeline in 1956. Although historians differ about the consequences of the pipeline debate, attempts by C.D. Howe, the powerful Liberal minister of trade and transport, to force rapid parliamentary approval angered many Canadian voters (Kilbourn 1970). An undercurrent of a new Canadian nationalism became redirected through dissatisfaction with political paternalism.

This emergent national pride was marshalled effectively by John Diefenbaker after his Progressive Conservative Party gained a minority government on 10 June 1957 (Robertson 2000).[3] The incumbent Liberals had held government for an uninterrupted twenty-two years, and the result was widely unanticipated. Diefenbaker was a small-town trial lawyer from Saskatchewan and was never really taken seriously by the central Canadian elites. Moreover, through his entire political career, Diefenbaker had very bad relations with the media (Brennan 1998). His personal relations were also difficult, and he reportedly had problems dealing with public servants (Rasmussen 1998). However, as a tremendous orator, Diefenbaker's political success was based upon public speaking tours across Canada. The economic and political decline of the United Kingdom, and the dwindling of export market prospects for Canada that this entailed, together with an increasingly diverse society based on post-1945 immigration, made Diefenbaker's arguments for new financial arrangements to mitigate growing regional disparities remarkably popular.

However, the prime minister was still in need of a significant mandate when the country went back to the polls only nine months after his first election. Diefenbaker's government was returned on 31 March 1958 with 208 seats – still the largest majority in Canadian federal history. Historian Ramsay Cook thus terms Diefenbaker's re-election "the most resounding victory in Canadian electoral history" (2005, 3). More remarkably, as well as taking 53.6 per cent of the popular vote, the Progressive Conservatives won 50 of the 75 available seats in Quebec.[4]

The Northern Vision and the New National Policy

In order to understand the need for this apparent detour into Canadian political history, it must be stressed that this victory was due in no small part to Diefenbaker's oratorical presentation of his Northern Vision. This "Vision for Canada" focused on pan-Canadian unity stimulated through government intervention and economic development of the North. Believing, like many earlier northern nationalists, that Canadians could be drawn together despite regional differences, the Diefenbaker government drew on the traditions of the Progressive Conservatives, specifically Sir John A. MacDonald's development of western Canada, and based an electoral campaign on a New National Policy. This policy coupled a vision of northern development with that of national service.[5]

In his opening campaign speech, "A New Vision," at the Winnipeg Civic Auditorium on 12 February 1958, Diefenbaker demanded support for this national development program. As well as the echoes of

Tennyson, the rhetorical connections established between the Arctic, scientific research, and Canadian unity are manifest.[6] Historians of the Conservative administrations of 1957–63 have often focused on the cancellation of the Avro Arrow fighter-interceptor project,[7] and the government's ignominious collapse following widespread Cabinet divisions over continental defence policy, together with the US-induced fall of the Canadian dollar (Gloin 1998; Simpson 1998). One analyst of Canadian public policy even argues that Diefenbaker's Northern Vision "was largely a creation of the northern bureaucracy attempting to aid its minister in an election year, at the same time selling their particular cause to the new government" (Nixon 1987, 292). Whatever the provenance of the vision, there has been a lack of historical investigation of the *scientific* policies that together formed the New National Policy. These initiatives were focused mainly on the *High* Arctic, and most notable among them was the founding of the Polar Continental Shelf Project. Furthermore, there has been little consideration of the impact of the vision on the scientists and public servants tasked with its implementation. Scholarship on government policy on northern Canada has tended to focus upon the Second World War and its immediate aftermath (e.g., Grant 1988).[8] However, there was an episode in Canadian history when young scientists were attracted into national service at the "nation's utmost bounds" in the High Arctic.

It is also important that later developments in the constitution of the national polity have meant that the brief period in which there was truly transcontinental support for this new sort of Canadian ambition has been almost completely forgotten. Diefenbaker appropriated interests in northern Canada common under the previous Liberal government for his own electoral success,[9] but what was unprecedented was the connection of government intervention, scientific investigation of the Arctic, and pan-Canadian nationalism. This construction of Canadian identity has always foundered upon the question of French Canada and the central place that this holds in Confederation. However, it remains the case that Diefenbaker won an unprecedented fifty federal seats in Quebec in 1958. As we will see in discussions of quotidian practices of science at PCSP Resolute, this historically limited form of pan-Canadian nationalism had legacies for organizations like the PCSP and the cultures of Arctic field science.

Alvin Hamilton, as minister of northern affairs and national resources, eventually brought the Action for a National Development Program to Cabinet in November 1958. As the new National Policy aimed to remedy the disparity of regional economies, much of the vision was focused upon the respective norths of the Canadian provinces. These northern areas were viewed as a vast resource hinterland, and the policy consequently attempted to develop transport and communications to improve access to mineral reserves, through what was known as the Roads to Resources initiative (Robertson 2000).

However, the delay until later in the year of discussion of this development program was due to the importance of one of the few *scientific* undertakings established by Diefenbaker's Cabinet, that is, the Polar Continental Shelf Project. A set of political developments had resulted in the interests of the Diefenbaker government shifting to the *High* Arctic by the spring of 1958. With the ink still wet on the ballot papers, the attention of the newly mandated government was being directed further north.

Hamilton, champion of the Roads to Resources campaign, had presented a memorandum to Cabinet that drew attention to USSR and US activities in the Polar Basin only three days after the election.[10] As Hamilton noted, Russian scientists had "drawn attention to the fact that the continental slope which bounds the deep-water Arctic Basin on the Canadian Arctic Archipelago side has been very imperfectly explored, and have expressed the opinion that it is a matter of great scientific interest to investigate the topography of this least-studied region."[11] According to Hamilton, following the *Sputnik* launches, the United States was also demanding rapid growth in Arctic research, "particularly on defence grounds."[12]

It needs to be remembered that these moves occurred despite attempts to integrate continental air defence arrangements. It is perhaps ironic that one of Diefenbaker's first acts as prime minister was to sign the North American Air Defence Command (NORAD) agreement on 1 August 1957, which moved towards integration of Canadian military capabilities under US control. The Canadian Arctic had massively increased in strategic importance as the development of long-distance ballistic missiles had made the Distant Early Warning (DEW) Line for bomber detection and interception effectively obsolete upon construction.[13]

The introduction of ICBMs initiated a transformation in cartographic practices, as there was a need for the widespread collection of gravitational observations on the ground. During the Second World War, it had become apparent that the national geodetic systems produced by different state mapping agencies did not correlate (Cloud 2000). Moreover, the various cartographic datasets were incoherent across ocean basins. This was problematic, to say the least, because missile weapons systems were completely dependent on accurate spatial positioning of points on the surface of the earth. As the mass of the earth is unevenly distributed, any given point needs to be positioned on the horizontal and on the vertical. The accepted method for determining a precise vertical position on the surface of the earth involves the mathematical formula, devised by Sir George Gabriel Stokes (1819–1903), for calculating geoidal undulation. However, in order to solve the Stokes function, gravity data are needed for numerous points distributed across the surface of the earth. Collecting such data for certain regions of the earth, particularly over oceans and at the poles, was both difficult and prohibitively expensive.

By the late 1950s, these requirements for gravity measurements for oceanic points could be met by submarines. US submarines were used to conduct gravity surveys through Hudson Bay during the summer of 1957, without the consent or prior knowledge of the Canadian research or political communities. As a member of the early PCSP field parties, and later director, George Hobson puts it, "There were a few people in Canada that were upset about that."[14] Arctic defence concerns began to be exacerbated by this growth in potential for submarine warfare, following the first transit of the Polar Basin by the USS *Nautilus* in 1958.[15] As the United States began to request geomagnetic and geodetic data for the most northerly regions of Canada, the Diefenbaker government became increasingly concerned about the sovereignty implications of *accompanying* demands for the complete integration of North American scientific research.[16]

During the IGY program, both the USSR and the United States also conducted geophysical research from a series of drifting stations established on ice floes in the vicinity of the North Pole (figure 1.1). The

1.1 | PCSP aerial photograph of Ward Hunt Island, 13 June 1962

Geophysics Research Directorate of the US Air Force Cambridge Research Center had conducted fieldwork based upon drifting ice stations since the early 1950s (Reed 1958; Althoff 2007). This rapid growth of the American and Soviet scientific presence in the Arctic, then, had produced anxieties over Canadian sovereignty.

However, Diefenbaker's "One Canada" mandate prompted determination to reconfigure geopolitical relations in the North American Arctic. Indeed, Hamilton directly referred to the prime minister's Winnipeg address when urging his Cabinet colleagues to recognize that expansion of scientific activity in the region was inevitable, but that it "would be most unfortunate to have it filled on a basis which would give the impression that Canada was incapable of doing research in the Arctic areas of immediate interest to us."[17] It was especially embarrassing for a government swept to power on a vision of the development of northern Canada to be in a position of potentially contested sovereignty.[18] The closest that Canada had to maps of the continental shelf had been obtained from the USSR.[19] The Diefenbaker government did not want to tolerate this situation.

This was all the more important as the First United Nations Conference on the Law of the Sea (UNCLOS I) in Geneva, 1958, had concluded that maritime states should have control over the resources on their continental shelves. A continental shelf is the submerged portion of a land mass, and the main Canadian concern was for the area to the north and west of the Queen Elizabeth Islands, under the relatively shallow water out into the Arctic Ocean. Canadian representatives at UNCLOS I had lobbied successfully for a convention on continental shelves that gave states natural resource rights over land covered by up to 200 metres of water and that also declared that straight lines from headland to headland, or the "sector" principle, would delimit territory.[20] States were furthermore granted the rights to resources in deeper offshore waters, assuming that it would be possible to exploit them safely. The convention thus greatly increased the potential territorial extent of Canadian sovereign rights. Moreover, field research by the Geological Survey of Canada had raised prospects of petroleum exploration in the region. It was thought by the Cabinet that, regardless of the scientific and political importance of Canadian research, there was also the very real possibility of deposits being found on the offshore shelf.

There had been discussions about the need for better co-ordination of Canadian Arctic research for at least the previous five years within the government's Advisory Committee on Northern Development.[21] It was therefore resolved, given the shift in geopolitical contexts, to initiate an interdisciplinary program of geophysical, oceanographic, hydrographic, and biological research on the continental shelf.[22] The foundation of this Polar Continental Shelf Project was recommended in March 1958 by a Technical Sub-Committee of the Advisory Committee on Northern Development, and significantly comprised representatives of various agencies and departments of the federal government of Canada.

The committee recommended that the PCSP should be an autonomous organization, separate from the requirements of other es-

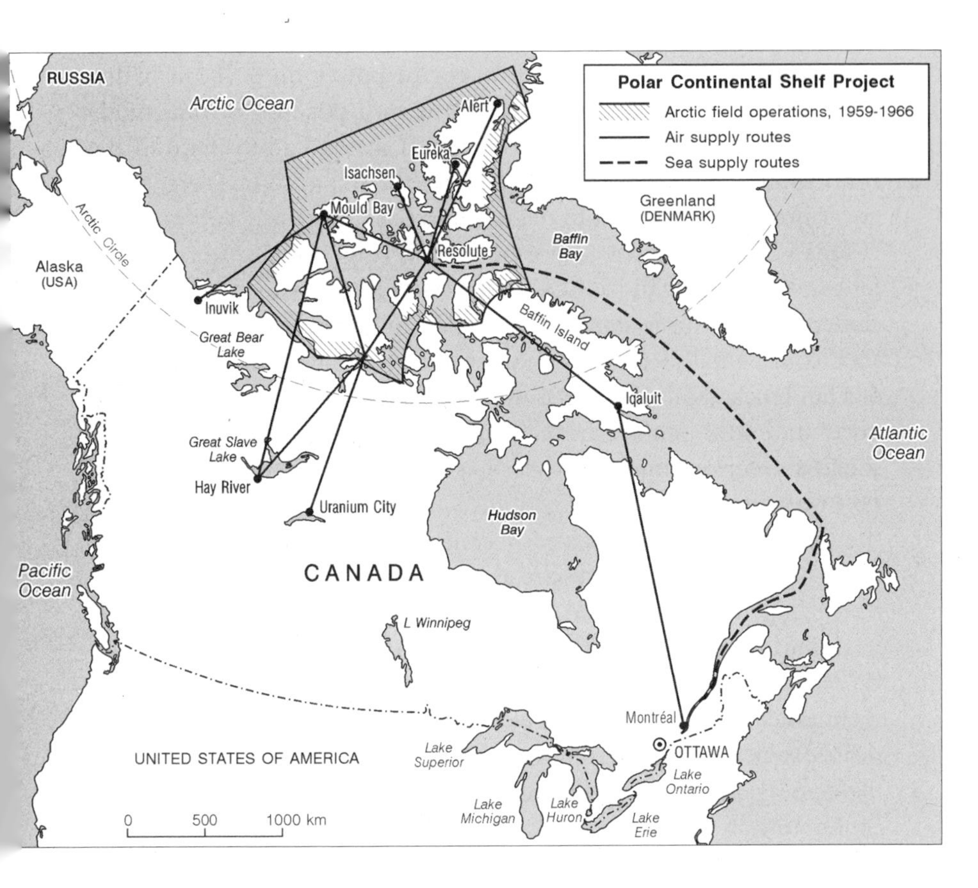

1.2 | PCSP Arctic Field Operations, 1959–1966

tablished agencies conducting Arctic research, such as the Defence Research Board and the Geological Survey of Canada (Roots 1960, 1962). As a result, the PCSP was created by a decision of the Privy Council at a Cabinet meeting on 22 May 1958.[23] Following fierce debates in Cabinet about which department would house this new research institution,[24] the PCSP was placed in the Department of Mines and Technical Surveys under the management of the director-general of scientific services, W.E. van Steenburgh. Preliminary logistical arrangements and a provisional field program were planned during the summer of 1958. By October 1958, E.F. (Fred) Roots, a British Columbian employed by the Geological Survey of Canada, had been appointed as the first co-ordinator of the new PCSP.[25]

As the PCSP was established during the IGY, most of the scientists involved had participated, in some way, in both programs. As Paul Comtois, minister of mines and technical surveys, noted for Cabinet discussion in May 1958, any PCSP staff member would be required to participate in the analysis of the results from the existing IGY program before any new fieldwork could begin in spring 1959.[26]

The first independent PCSP field party began survey and reconnaissance work in March 1959. A full-scale field season was to commence during the following spring of 1960. As will be discussed further in chapter 3, by the mid-1960s, PCSP research had grown to cover the entire Canadian High Arctic (figure 1.2). The majority of field scientists were seconded to PCSP from other branches of the federal government, such as the Geographical Branch or the Canadian Hydrographic Service. The underlying philosophy was therefore that, for any given scientific branch of the government of Canada, PCSP provided "logistic support to enable the Branch to carry out its own programme in that particular subject, in an area where it would otherwise be expensive or impossible to operate; and it supplies the scientific coordination between this study and related studies."[27] The PCSP was to *synchronize* scientific research in the Arctic, because the difficult and expensive conditions might otherwise prove prohibitive to fieldwork.

Constructing Scientific Sovereignty?

Given the sensitive nature of their field sites, PCSP scientists were soon drawn into discussions about the geopolitics of the Arctic. Whilst managing the PCSP, van Steenburgh was to write, in response to a memorandum on Canadian sovereignty, "We feel that we must be realistic in asserting our sovereignty in the Arctic regions. The increasing interest and activity of the USSR and the USA in the Arctic Basin will have a direct bearing on any decision affecting our claims. With these considerations in mind, it would appear that the sector approach to sovereignty in the Arctic Basin is unrealistic and unacceptable to other interested nations. Canadian claims must be based on firm logical reasoning."[28]

The PCSP, because of the nature of the topographic, cartographic, and geophysical knowledge that it collected, had a critical role in deciphering and depicting sovereignty over constantly changing ice regimes. The natures of ice, for van Steenburgh, raised questions about depicting territorial sovereignty in the region that only the PCSP could answer:

> I feel that good sound reasons can be advanced for sovereignty to the Arctic Archipelago, the adjacent islands, the surface of the fixed ice, and the resources of the Arctic Continental shelf. Beyond the line of fixed ice presents a more difficult problem … I might add that we will have much more precise information on the Arctic Basin conditions following the Polar Continental Shelf Project, which will start in March 1959. At the moment precise descriptions and definitions are difficult to produce. Canadians must do more exploratory and development work in the Arctic regions or international; questions will be raised regarding any area over which we claim sovereignty.[29]

Furthermore, Fred Roots appears always to have held suspicions about sovereignty anxieties intruding into the quotidian business of science. Shortly after being appointed to the PCSP, Roots responded bluntly to debates within government about Canada's political jurisdiction, given concerns about drifting ice stations in the Arctic:

> For the most part, the scientific work done on the ice islands is available to the world at large. It is highly probable that the value of these studies would be lessened, and the programs of study curtailed or planned differently, if there is a danger that every time the ice island drifted past a certain geographical position it changed sovereignty … In addition to limiting the "carrying out" of scientific investigations, I would add that there might be restrictions on the planning and financing of such activities, and on the interchange of information. If the area were divided politically, the work done in each sector would probably become more of an immediate, strategic and military nature.[30]

For Roots, the spectre of sovereignty disputes could seriously impede progress on scientific questions. This reluctance was in further evidence during interviews with Roots, as if acknowledging the *territorial* imperative behind the founding of PCSP would somehow

1.3 | PCSP survey party at "hill" of granitic boulders on Arctic Ocean, 20 km northwest of Cape Isachsen, Ellef Ringnes Island, 24 April 1959

devalue its *scientific* achievements.[31] Both dimensions were addressed during these early field seasons (figure 1.3). As I will discuss in chapter 2, this betrays a long-standing tension between expeditionary praxis and scientific activities in the Arctic. Moreover, such controversies might allow excessive military influence upon scientific investigations.

The reasons given by the Government of Canada for founding PCSP revolved around the importance of the field research for global scientific issues and national interests during the IGY. However, the importance of the IGY was not obvious amongst all the participants involved in the scientific fieldwork. Rather, as Denis St-Onge, a geomorphologist seconded onto the first field parties in 1959, argues, sovereignty disputes continually circulated through scientific practices.

> You know, we had heard about the International Geophysical Year, but it was … If there was a link it certainly was not obvious to people like me at the time … They were making these little links all over the place, but the real force behind this … [laughs] … The official line was in fact that … Canada needed to know more about its continental shelf. Hence the name, which really makes very little sense, because a lot more scientists were

> working, and ended up working, on the land and all sorts of other things rather than the shelf. The shelf was, became, incidental practically from the start. But the real reason behind all of this is that Canada needed to assert its sovereignty. That is the real reason. And the people realized very quickly that one way of doing this somewhat inexpensively is to send a bunch of scientists up there every summer. It doesn't cost very much. They'll work because they're dedicated and this will help serve Canadian sovereignty. But the Canadian sovereignty was always paramount, although it was never articulated. But that's the official reason. There's no doubt about that whatsoever.[32]

As will be discussed further in chapter 5, Canadian sovereignty has been a recurrent issue of contestation across different communities in the Arctic. But what is important to stress here is that, for the young Canadian inspired by Diefenbaker to adventure at the nation's bounds, there was nothing better than work literally to redefine the extent of "Canada." Even in his castigation of adventurous activities in the Arctic, the memories of Arthur Collin, a graduate student at the University of Washington (Seattle) when he joined the PCSP in the field, are revealing:

> When we started, there was a real sense of urgency. There was a sense of not only national urgency, but there was a sense of international urgency as well. And this is why it could attract the people that it did, because we all felt that there was a real international thrust. There was a real international sense of contribution that we could make. And all of us had that sense, within our disciplines and within our sense of *nation* … All my colleagues felt that it was a national urgency. And we all recognized that it was an extraordinary personal opportunity as well … Even if you were just up there as an adventurer, and there were some. Those of us that were interested in the science recognized that it was an extraordinary opportunity, because just getting there is such a formidable undertaking, you know. I spent many days on the Arctic Ocean … Many days? Months! Which I would never have had the chance to do before. So, we all recognized this in the back of our minds. There was a sense of responsibility as

> well, which was interesting, because we were all very young men, and we were working very hard, many hours. We often worked night and day because of the logistics.[33]

Denis St-Onge makes a similar point: "It is true that the first two or three years, we knew that we were pioneers in many ways, we were going in an area that was little known, and we were in a position where we should be able to contribute something new to knowledge in general and to this part of Canada. So there was no doubt. It's not something you discussed a great deal but, but we all knew this. And as a result there was a sort of dedication to the work that resulted from that … You did whatever was required."[34]

It becomes evident, then, that Mertonian ideals for scientific communalism,[35] political motivations for sovereignty, and romantic constructions of Canadian nationalism were brought together through the quotidian practices of PCSP fieldwork.

Although the IGY inspired the ideal of a Canadian organization undertaking field science, it was the legacy of Diefenbaker's vision that was to prove more pervasive. George Hobson, the leader of the first PCSP seismic field party in 1960 and, from 1972, the second director of the project, recalls that Diefenbaker still remembered the PCSP during the late 1970s: "There was a half-hour documentary [about PCSP] made by CBC [Canadian Broadcasting Corporation]. And evidently Diefenbaker saw it. I had a call at the office the next morning and he wanted to know if that was the same Polar Continental Shelf Project that was put in place while he was prime minister."[36]

Despite all his later difficulties and the failure of the notion of pan-Canadian nationalism during later shifts towards a multiculturalism that recognized the rights of the Québécois and indigenous peoples, Diefenbaker was to remain faithful to the vision that he had outlined. And, as we will see, it was this vision, and the desire to reassert Canadian sovereignty that lay behind it, that continues to pervade the contemporary activities of the PCSP.

Conclusion

This chapter has examined the legacies of cultural understandings of nordicity for scientific practices in the post-war Canadian Arctic. By considering developments in High Arctic scientific research under the New National Policy of the Progressive Conservative governments of 1957–63, it has been argued that histories of science must be set within the narratives of national polities.

The geopolitical shifts that *Sputnik* initiated had ramifications for both Canadian electoral politics and scientific practice. The thirst for geographical and gravitational knowledge about the northern regions of Canada threatened territorial sovereignty, as well as revivifying dormant national identities. It was the appeal to this romantic mindset that encouraged many scientists to participate in Arctic fieldwork. The attempts to inspire national unity in Canada circulated through the organization of field science in the Arctic.

Furthermore, I have argued that attempts to create a Canadian century were difficult to meld with attempts to establish scientific credibility among Canadian field researchers. Although desires to bring modernity to the entire land mass of Canada had widespread popularity in 1958, these moves were often impeded in practice by the new sense of independence felt by young Canadians – an independence that was inspired by the New Vision in the first place.

It is unsurprising that these stories have been hidden for so long, given the security restrictions on much archival material. By recording some of these histories, I hope to have indicated that demands to meet scientific goals and establish Canadian sovereignty were in constant tension. The anxiety at the heart of the IGY was thus replicated within PCSP, an institution that was inspired by similar Mertonian ideals of interdisciplinarity and multinational cooperation. But what is important about these narratives of the Polar Continental Shelf Project is that they signify an episode in Canadian history when young scientists were attracted into public service at the "nation's utmost bounds" in the High Arctic. It should not be forgotten that Diefenbaker's vision resulted in many legacies.

2 Between Observation and Experiment in Arctic Fieldwork

"Suppose you tell me what you know," he [Lord Boreal] said.
"They're doing experiments in the North," Lyra said.
Philip Pullman, 1995, *Northern Lights*, 95

Introduction

Philip Pullman's novel, *Northern Lights*, concerns a young heroine, Lyra, who is drawn impulsively from her home in Oxford all the way to the Polar Sea in order to investigate wicked "experiments" being conducted upon kidnapped English children. It turns out, in this particular narrative, that experimental surgery on young children allows the creation of an entry bridge to another universe through the aurora. As discussed in chapter 1, such constructions of the North as locales of adventure are a trope in northern literature. It is worth remarking that adventurous discourse regarding the North has always been concerned with demonstrating the conquest of nature. A nature that is exactly *in*temperate. In this chapter, I will argue that the idea of a *northern experiment* has held even greater epistemic cachet precisely because it involves *control* over those pathological natures.

In what follows, I illuminate this relationship between the Arctic as an *expeditionary space* and the Arctic as an *experimental space*. I will show that an elision goes much deeper than children's fiction and even entered into published discussions about the best conduct of field science during the late 1950s and 1960s in the Canadian Arctic. By examining the practices of environmental scientists supported by the PCSP, I argue that what Bruce Hevly has in a nineteenth-century, European-Alpine context termed the "authority of adventurous observation" was still evident in this later period (Hevly 1996, 68). This is notwithstanding the best efforts of Canadian field scientists to deploy a *precarious authority of experiment.*

The sociologist of science Andrew Pickering has written of the *mangle of practice* that occurs during quotidian activity by scientists (Pickering 1995). Through this, Pickering suggests that "the contours of material and social agency are mangled in practice, meaning emergently transformed and delineated in the dialectic of resistance and accommodation" (23). In the field sciences, scientific practice often mangles epistemic distinctions between experiment and observational measurement. As Robert Kohler states, field scientists are used to adapting to "a world that takes for granted that experiments are the better, or even the only, way of knowing nature" (Kohler 2002a, 1). This aspect of field practice will become evident, specifically, in the attempts by PCSP scientists to *adapt* experimental methods to field situations. In examining some cases of Arctic fieldwork, this chapter will investigate how the misconstrued importance attributed to successful field experiments resulted in physical, epistemic, and ultimately emotional difficulties for PCSP scientists.

The Practices of the Geographical Sciences in Arctic Environments

The emergence of spatial thinking across the social sciences and humanities over the past two decades affected historians and sociologists of science, who began to emphasize the *placing* of knowledge claims (Ophir and Shapin 1991; Smith and Agar 1998). Geographers of science, developing these insights, have drawn attention to the importance of space and place in the construction of scientific knowledges (Livingstone 2003; Powell 2007a; Withers 2001, 2009).

Such discussions have generally succeeded in demonstrating that scientific practices can be understood *geographically.* However, these revivified historical geographies of knowledge have been much less successful in influencing the historiography of what have been understood as the "geographical sciences." The complicated diversity of topographical, observational, and experimental practices involved in the geographical sciences, and its relative implications for practitioners, has often been neglected. The complexity of the geographical and environmental sciences have only recently begun to enjoy attention from historians and sociologists (Braun 2000; Oreskes and Fleming 2000; Doel 2003; Miller 2004; Doel and Harper 2006). However, as Mott Greene complains, there is still sufficient lack of interest such that the "history of the geosciences will probably be written, for some time yet, mostly by earth scientists themselves" (1989, 331). The work of historian Naomi Oreskes (1996, 1999) on the twentieth-century history of geology is, of course, an important exception.

Studies of the geographical sciences have demonstrated the imbrications between actual practices and the desires of the defence industries in the Cold War period (Jones-Imhotep 2000; Cloud 2000, 2001a, 2001b; Oreskes 2003; Dennis 2006; Doel, Levin, and Marker 2006). As historian Michael Dennis points out, "It is no longer sufficient to ask if military patronage affected the production of knowledge; instead we need to examine particular knowledge claims and the context in which they are made" (2003, 817). Historians of geography have also started to provide fine-grained analyses of the geopolitics of science during the Cold War (MacDonald 2006a, 2006b; Barnes and Farish 2006). As geographer John Cloud argues, "There was no either/or dichotomy[,] … most participants devised ways to accomplish both" basic research and military objectives (2003, 630). Moreover, as was discussed in Chapter 1, geographical knowledges were central to developments in geophysical science. The emergence of new military technologies, such as ICBMs and nuclear-powered submarines, effected changes in global geographies such that any site could be rendered *strategic* (Cloud 2001b).

During the 1950s and 1960s, the polar regions, as new targets for both geographic knowledge and military strategy, became enveloped with such developments in the geophysical sciences. Polar scientists, because of their proximity to international networks and expertise regarding strategically sensitive regions, were "swiftly incorporated into highest-level discussions about the relationship of science and foreign policy" (Doel 2003, 646).

Historical studies that have emerged tend to focus on the connections between the geosciences, geopolitical strategy, and the military. The neglect of the epistemic and emotional practices of the geographical sciences is important in the polar regions, because the connections between scientific and exploratory praxis have been significantly intertwined. Cultural historian Max Jones (2003), for example, has shown how scholars have often disregarded, or even worse, scoffed at, the important scientific *intentions*, under the auspices of the Royal Geographical Society, of Scott's expedition to Antarctica. This was no less the case for PCSP scientists.

Moreover, recent histories of cartography have stressed the need to take account of *particular* practices of surveying in different regions (Edney 1997; Burnett 2000; Craib 2004). The environment of the Arctic has provided obstacles to "description, measurement, charting, and even physical penetration" by non-indigenous peoples (Collis 1996, 27). The geographical environment thus meant that survey of the Canadian Arctic was difficult, resulting in the adoption of specific practices.

Furthermore, I argue that the relation between geographical *survey* and geographical *science* in the polar regions has not been fully understood by historians of the geosciences. The distinction was, for PCSP fieldworkers at least, crucial.[1] Although political geographer Klaus Dodds is right to stress that the "collection of accurate geographical information was essential" to justify Britain's claims to its South Atlantic territories, it is misguided to transfer this to a wider generalization across all polar environments (2002, 24). The actual geographical practices that were to be deployed in the service of territorial sovereignty were in flux by the early 1960s, and this was no less the case in High Arctic Canada. As I will show, it is far too simplistic to argue for the High Arctic, like Dodds is able to do for Antarctica in this period, that "scientists were beginning to replace explorers" (Dodds 1997, 56).

This chapter examines an epoch when, in the Canadian Arctic, geographical scientists moved from a conception of good scientific activity as involving topographical survey, to one of structured observational measurements, and then to field experiments, in a period of around fifteen years. In the pages that follow, I reveal some of the complicated relationships between activities that were construed to be scientific whilst also allowing some sort of demonstration of territorial sovereignty. However, before considering how these debates affected the PCSP, it is first necessary to understand what a field experiment was taken to be in the 1960s, and why it was deemed to be so important.

The Philosophy of Experiment

A number of studies by historians and philosophers of science have remarked on the need for scholars to study *actual* experiments (Hacking 1983; Shapin and Schaffer 1985; Gooding, Pinch, and Schaffer 1989). As historians remark, "Experiment is a respected but neglected activity" involving "an active process of argument and persuasion" (Gooding, Pinch, and Schaffer 1989, xiii, xvi).

This is an important injunction. Experimental practices are far from trivial. However, it has led to a general lack of attention from students of science studies to why scientists *believed* that they should undertake certain kinds of actual experiments. Bruno Latour entirely collapses the distinction made between observational measurements, whereby phenomena over which humans have no control are *recorded*, and experiments in the field, whereby phenomena are (supposed to be) *manipulated* by humans usually (though not necessarily) using instrumentation, because the results *brought back* from the field from these two procedures appear visually alike (Latour 1990). This epistemic reduction has been very influential in the readings of experiment by geographers. Many historical geographers of science would seem to concur with Scott Kirsch when, in a study of the deployment of US nuclear technologies in the 1960s, he argues that the observational-experimental distinction is merely semantic (Kirsch 1998).

Rather than become entwined within an argument about theories of experiment, I prefer to draw from ethnography. It is important to state that for PCSP scientists this observational-experimental distinction was absolutely critical to their *sense of self*. In order to understand this point, it is necessary to get a better handle on what an experiment was supposed to be in the 1960s. I should stress again that the discussion of the meaning and purpose of field experiment in this chapter is limited to the environmental field sciences between 1958 and 1970 in the

Canadian High Arctic. Different understandings of experiment are, of course, distributed spatially, temporally, and epistemically across scientific pursuits.

However, for Karl Popper writing in 1957, scientists should attempt to emulate physicists, and "physics uses the method of experiment; that is, it introduces artificial controls, artificial isolation, and thereby ensures the reproduction of similar conditions" ([1957] 2002, 7). These requirements for control and replication were fundamental to post-war philosophies of science.

The philosopher Rom Harré provides a clear examination of the supposed distinction between structured observational measurement, or "'an exploration,'" and an experiment (Harré 1981, 23). Explorations are important, for Harré, forming "a kind of intervention in the natural world which yields knowledge, but lacks the manipulative character of the true experiment" (23). An experimenter, Harré argues,

> is in a different relation to natural things. He actively intervenes in the course of nature … Experimenters describe their activities in terms of the separation and manipulation of dependent and independent variables. The independent variable is the factor in the set-up that the experimenter manipulates directly. The dependent variable is the attribute which is affected by changes in the independent variable … By careful design of an experiment it is possible to maintain constant all properties except those one wishes to study, the dependent and independent variables. A property which is fixed in this way is called a "parameter." Fixing the parameters defines the state of the system within which the variables act. (22)

A consequence of this requirement for the separation of variables and fixing of parameters, as Harré himself states, is that it "seriously restricts the use to which experiments can be put" (1981, 22). Many phenomena therefore are simply not practically amenable to experimentation. However, such practicalities did not dampen the enthusiasm for experimental methods amongst many environmental scientists in the post-war decades.

During the 1960s, despite the obvious derivation of such models of explanation from analyses of laboratory physics, various philosophers of science, including David Harvey in *Explanation in Geography*, termed this approach to experimentation "the '*standard*' model of scientific explanation," under which any hypothesis with a significant degree of confirmation was a scientific law (1969, 30; *my emphasis*). Moreover, under this system, the definition of progress in scientific activity relies on the establishment of such laws, which normally proceed through the identification and confirmation of physical processes under laboratory experimentation.

Notwithstanding the popularity of this model, philosophers have begun to argue that the situation is very different in the environmental sciences because they involve *complex* systems (Oreskes and Fleming 2000). These are systems in which all variables vary, and different mechanisms interact continuously, thus making the drawing of boundaries to effect some sort of experimental closure highly imperfect. As geomorphologists Michael Church and David Mark put it, rather than from deductive reasoning and theories, "most inquiries, at least in the field sciences, are initially prompted by the observation of formal relations" (1980, 342).

However, necessary statements in theories, as the apogee of explanation in this scientific model based on physics, were to be achieved through controlled experimentation. And a *scientific* experiment was thus designated, to quote a particularly succinct definition from a later paper by Church, in the following way: "A scientific experiment is an operation designed to discover some principle or natural effect, or to establish or controvert it once discovered. It differs from casual observation in that the phenomena observed are, to a greater or lesser extent, controlled by human agency, and from systematically structured observations in that the results must bear on the existence or verity of some conceptual generalization about the phenomena" (Church 1984, 563).

The requirements for a classical experiment are thus constructed upon a set of social practices established for a laboratory space. It should be evident, to say the least, that this sort of employment of experimental method is rare in the environmental sciences. As well as the relevant variables being covariant over space-time, complex, open systems are often too large, and the temporal scales of interest too long, for convenient experimentation to test theories. Moreover, phenomena of interest are often too public and the costs prohibitively high for legitimately designed experiments. Indeed, any open system reduced to elements amenable for experimentation could be construed as inherently *dysfunctional*, because in attaining analyzable simplicity the variation that the system depends on for stability is removed (Church 1984). At the same time, however, the very scientific status of the geographical sciences, such as hydrology, glaciology, and geomorphology was viewed to depend upon their ability to establish such a "corpus of rational statements" (Church and Mark 1980, 381; Church et al. 1985). Otherwise such activities, as geographer Barbara Kennedy argued nearly forty years ago, would remain at a natural history stage of scientific progress by relying on collection, description, and classification of observations (Kennedy 1977, 1979).[2] Herein lies the dilemma: how to master the epistemic resources of experimentation for environmental systems? How might experimental practice be mangled for the field?

Field Experiments in the Environmental Sciences

Despite the limitations discussed, laboratory experiments are still often conducted in environmental science, but inevitably must either involve exceptionally short spatio-temporal scales, or scale models that are susceptible to scaling effects. Precisely because of these restraints, successful experimentation in the space of the field was deemed to be crucial to establishing theories for the field sciences (Church 1984).

The concept of the field experiment as deployed by environmental scientists in the 1960s had first emerged in agricultural field trials in the 1930s (Yates 1970). Developments in statistical techniques after 1945, coupled with increasing sophistication in the understanding of experimental design, facilitated the widespread adoption of field experiments in the environmental sciences (Fisher 1960). As such, experiments were designed for situations "in which the effects under investigation tend to be masked by fluctuations outside the experimenter's control" (Cox 1958, 1). Such fluctuations are usually meteorological in the environmental field sciences.

By employing principles of experimental design *sensu lato*, "comparative experiments" could proceed in the field, whereby a number of alternative treatments are applied to individual "experimental units," and observations (or a set of observations) are made for each unit (Cox 1958, 4, 12). As Cox argues in the standard textbook used by post-war field scientists, the object is thus "to be able to separate out differences between the treatments from the uncontrolled variation that is assumed to be present" (12).

Other types of experimental procedures have been adopted by field scientists (Henke 2000; Rees 2001; Arpin and Granjou 2015). For Kohler, such "practices of place" have been developed in order to maintain the credibility of the field sciences (2002a, 213). Kohler expands this point as follows: "Precise measurement of environmental variables was more place dependent, but though many lab instruments were too fragile and precise for field use, they could be adapted, and suitable places found for their deployment. But experiment was different: this quintessential laboratory practice loses its power outside the controlled and placeless place for which it was designed" (2002b, 195).

Such practices may also, for Kohler, include attempts by fieldworkers simply to find field sites where a single environmental variable changes while others are constant (2002a). In this way, Kohler argues, waiting for "the appropriate time and the right place to observe in nature is no less active than fixing conditions in a lab" (2002b, 200). It is by developing these "skills of selecting and using places" that scientists are able "to operate effectively in the field" (205).

However, as experimental design was quickly adopted by environmental scientists in the 1960s, many field scientists were much less conservative than Kohler admits in their attempts to draw from experimental methods (Kennedy 1992). The popular approbation of experimental methods became problematic, for some field scientists, when many began to designate any set of structured observational measurements as an experiment, neglecting the important component of controlled, anthropogenic modification of isolated variables (Ahnert 1980; Church 1984; Le Grand 1990). For many field scientists, this perceived sloppiness was directly attributable both to philosophical and statistical naivety, as well as the difficulties harsh environments imposed on "testing theory in the field with rigour" (Church and Slaymaker 1985, xi). In order that the repercussions of this epistemic confusion be fully comprehended, it is necessary to examine some particular cases of fieldwork.

For much of the twentieth century, the Arctic has been constructed as "a pristine natural laboratory for the field sciences" (Bravo and Sörlin 2002, vii). Moreover, this space was to allow the performance

2.1 | Fred Roots making a presentation, circa 1961

of in situ experiments. Such desires for successfully conducted field experiments in the Arctic were exacerbated in the immediate post-1945 sciences.[3] Indeed, experiments formed part of a much wider ambition in the post-war Anglo-American natural and social sciences to deploy hypothetico-deductive schemata. Thus by focusing on environmental sciences in the Arctic during a period when immense excitement was devoted to field experiments, we can begin to understand how the geography of science has, in David Livingstone's words, an impact on the "cognitive content" of science (2002, 89). In order to examine field experiments in High Arctic environmental science, I will focus on some of the work done by members of the Polar Continental Shelf Project.

Fred Roots and the Establishing of "Good Practice" in Arctic Science

As discussed in chapter 1, the PCSP was established in 1958 and launched an interdisciplinary program of geophysical, geographical, and biological studies on the continental shelf northwest of the Canadian Arctic Islands. E.F. (Fred) Roots was appointed as the first coordinator of the PCSP (figure 2.1). Roots was a geologist from British Columbia and had distinguished experience of multinational fieldwork in Antarctica.[4] Born in Salmon Arm in July 1923, Roots had spent much of his childhood as a mountaineer, and studied geology at the University of British Columbia, before taking his PhD at Princeton. After working for the Geological Survey of Canada (GSC) for six years, he was invited in 1949 by the Royal Geographical Society to join the Norwegian-British-Swedish Antarctic Expedition, 1949–52, as a Commonwealth representative. Roots earned a reputation from his Scandinavian and British colleagues as a quiet, modest, and extremely able geologist and "field person."[5]

Having had to resign from the GSC to join the expedition, Roots went to Cambridge to write up his results as a research fellow at the Scott Polar Research Institute and the Department of Mineralogy and Petrology.[6] He then taught geology for a year at Princeton in 1953, before returning to Cambridge until January 1955, when he accepted a new post with GSC.

By the summer of 1955, Roots had returned to Canada to run logistics for the GSC's Operation Franklin in the vicinity of the Queen Elizabeth Islands. This twenty-eight-person expedition demonstrated the utility of combining various disciplinary research agendas, surveying 260,000 square kilometres of little-studied terrain, and indicating the presence of thick accumulations of sedimentary rocks and structures (analogous to those found in oil fields) (Nassichuk 1987; van Steenburgh, Fortier, and Thornsteinsson 1964). This operation also indicated the successful deployment of modern technology in the field, especially in flight support by two Sikorsky S-55 helicopters.

Under the leadership of Roots, a PCSP field party began work to establish a surveyed baseline to facilitate an operational navigation system in 1959 based out of the Joint Arctic Weather Station at Isachsen, Ellef Ringnes Island, and the first full-scale field season commenced from Isachsen during the following spring of 1960. This area of initial interest, chosen for both reasons of convenience and the congruence of a number of channel entrances among the Arctic islands, was a band 300 miles long, centred on Isachsen, and stretching 100 miles out to sea.[7]

2.2 | Helicopter landing on an ice ridge, May 1971

It was exactly the overcoming of the traditional challenges of Arctic fieldwork that, for Fred Roots, defined the PCSP. In promotional literature produced by PCSP, the project was supposed to have run "down the curtain on the expeditionary 'hit-and-run' approach to arctic research" and replaced this with "a sustained, long-term, integrated effort" (Polar Continental Shelf Project 1974, 5).

The Physicality of Arctic Fieldwork

Over the course of its early years, the PCSP was able to establish a number of logistical precedents in Canadian Arctic research.[8] These included the establishment of a 900-watt low-frequency Decca "6f Lambda hyperbolic" survey and navigation system and the marking of mid-March to May as the best period for air support of parties on sea ice (figure 2.2). In doing so, by as early as 1959 Roots was able to claim to his superior in the Department of Mines and Technical Surveys,

2.3 | F.P. Hunt using a Tellurometer in field near Camp 200, 1970

W.E. van Steenburgh, that "the party has managed to do much work that some old Arctic hands had considered impossible" (figure 2.3).[9]

Moreover, as a consequence of its departmental autonomy, the PCSP was able to remain flexible, even footloose, not only in organization but also in combining interdisciplinary scientific research programs in unprecedented ways. As Roots wrote for a PCSP Steering Committee meeting in 1960, the basic principle was "to complete a thorough study of a problem, rather than general wide-ranging shallow surveys."[10] Roots was to reminisce in a later interview, "I was insistent from the beginning that whatever we did it would not be reconnaissance for its own sake: the work we were going to do in the High Arctic should be of the same scientific quality and accuracy as work done anywhere else, regardless of latitude" (Foster and Marino 1986, 16).

As I have argued, this emphasis on scientific practices over more ostensibly adventurous pursuits is a typical conceit of descriptions of Arctic activity. As Christy Collis remonstrates, many Arctic narratives deploy familiar patterns "in order to construct the North as Eden, as hell, as passively empty, as savagely adversarial, as picturesque, as sublime, as female, as godless, and as God's trial of man" (1996, 27–8). What was different in the PCSP case was that the progressive edge of science was to involve successful deployment of the *experimental method* in the field. In the PCSP's sea-ice studies, for example, Roots was keen to describe "a series of experiments … carried out to determine the effect of coating the surface of sea ice with various materials that changed the index of solar absorption."[11] These were complementary with a similar series of experiments conducted on Meighen Island icecap by Keith Arnold, a geographer seconded to PCSP from the Geographical Branch, using aluminium foil and chicken wire. For Arnold, the nature of these experimental practices should provide for great publicity for the PCSP, "being something that the public could grip on to right away."[12]

Scientists employed by PCSP thus intended to move beyond arguments about establishing physical presence in the Arctic by conducting structured observations and, ultimately, experiments. As Denis St-Onge, another member of the first PCSP field parties and later president of the Canadian Association of Geographers, argued, there was little tolerance in the field for the "Boy Scout attitude to Arctic fieldwork."[13]

In order to exorcise any hints of adventurous discourses to facilitate scientific progress, the PCSP scientist had to meet a number of challenges. As Sverker Sörlin has argued, the "'holistic hero'" of the nineteenth-century Arctic was always a "scientific omnivore," competent across a number of disciplines (2002, 109). This was no less the case with PCSP. However, participation in the PCSP recast the identity of the field scientist in a more *modern* idiom.

In the first place, the PCSP field scientist had to develop a corporeality that was dependent upon *environmental immunity*. The ability to employ instrumentation regardless of conditions was important to Roots in the composition of the field parties. In a discussion of the selection of field personnel, Roots remarked, "I wanted Frank Hunt. He was far and away the most experienced winter topographical surveyor we had at that time. He had run the 60th parallel survey in wintertime. He had already shown that just because it is cold and the wind is blowing, there is no reason to be less accurate with a theodolite than when it is nice and sunny" (Foster and Marino 1986, 25).

In the second place, PCSP scientists had to be competent managers of logistics. This should be clear after consideration of the objectives that the PCSP had to achieve. As Roots summarizes,

> We needed not only to go from place to place but to be able to know precisely where we were at all times and to return at different seasons of the year at different times with different equipment to the exact, precise, same spot. To do this we had to have a combination of surface transport and air transport equipment. If we used helicopters, ... we had to have two, in case one broke down. To fly a helicopter you had to have fuel, that meant fixed-wing aircraft that could land anywhere on the ice to carry the fuel ... In the end result, we found ourselves arranging for a small airforce with its own communications, its own navigation equipment, [and] its own supply route. (E.F. Roots interviewed in Energy, Mines and Resources Canada 1986)

Third, scientists in the field had to be skilled at various sorts of manual tasks, such as small-scale mechanical and electrical engineering. As a result of such responsibilities, St-Onge reminisced, "I was supposed to be there to do science; I ended up being mostly mechanic the first summer."[14]

Moreover, those in positions of perceived lower status within the PCSP were also expected to be multi-skilled in order to facilitate successful fieldwork. In interviews with PCSP mechanics and field support staff, this multitasking during long working days in difficult conditions was stressed above all other recollections.[15] In 1959, for example, an assistant cook was hired from Quebec because, despite his "modest" cooking abilities,[16] he was fluently bilingual, and "had a very wide background of practical experience in the mechanical, electrical, and woodworking trades, plus considerable 'bush' experience as a trapper and lumberman."[17] Despite deploying all these skills, this young man also took full responsibility for the kitchen at the Isachsen base camp for the entire season as well, as the chief cook failed to arrive for his duties in the field.[18] This stress placed on bilingualism and "bush experience" also reaffirms the vision of Canadian nationalism prevalent within the founding discussions of the formation of the PCSP.

The importance of being a competent fieldworker is evident in other discussions of field science in northern Canada. In a retrospective on the career of permafrost specialist J.R. Mackay in the western Canadian Arctic, a contributor argued that any "adventure" by a "good scientist" was supposed to be merely the result of "some miscalculation" (Mathews 1985, 8; Jahn 1985). However, the cultivation of such characteristics as "physical fitness" and "curiosity" by the field scientist allowed for the performance of successful experimentation (Mathews 1985). Adventure had to be overcome in order to practise science.

The Rigours of an Arctic Experiment

As might be expected, however, the quotidian enactment of such ideals regarding the reconfiguration of scientific practice and individuality by the PCSP was problematic. These difficulties were often encountered by the scientists precisely because, despite their experimental aspirations, the ultimate reason for the funding of the PCSP was the completion of the cartography of Canada in defence of national sovereignty. This was most evident in a Government of Canada press release that Alvin Hamilton had demanded as part of increasing public awareness of northern Canada, and was issued as the first PCSP field season commenced in 1959.

> Starting this spring, Canada is going to send scientific expeditions each year into the forbidding arctic wastes north of the Canadian archipelago and beyond to carry out a broad program of research on the rim of the Basin … The project, when fully organized and operating, will be the biggest and boldest scientific venture of its kind ever undertaken in northern Canada. Scientist-adventurers of rugged stock are needed to man the expeditions: men imbued not only with love of research but with the same venturesome spirit that sent men like Hudson, Frobisher, Baffin, and Davis into the trackless wastes of the Arctic to probe its secrets.[19]

It was precisely this ambivalence, prevalent in the identity of the "scientist-adventurer," that made Roots's goals for Arctic field practice so difficult to achieve. PCSP literature would commonly refer to the fact that the "men involved … are conscious of making history" or that some employees were "the first human beings to set foot on some of the Arctic Islands, hitherto known to geographers only from air photographs" (Polar Continental Shelf Project 1974, 6). Those involved in PCSP fieldwork thus clearly constructed themselves as different from those "ordinary 'other scientists'" working further south (Polar Continental Shelf Project 1974, 3).[20] The everyday performance of field practices became a quotidian arena of contestation between science and adventure.

Indeed, Roots himself was to argue in print towards the end of the 1960s that logistical difficulties in consequence of the environment of the High Arctic had an immense impact on the research conducted (Roots 1969). Such could be the impediment to research, argued Roots, that there are "many investigations whose subject and priorities have changed markedly because of logistics considerations, and others where the investigators have become so dominated by the logistics challenge that they have been compelled to list as 'results' the number of miles travelled or samples taken, rather than show the increase in understanding of the subject" (68).

A PCSP publication was even to state that this situation informed "an entire psychological theory of arctic research" articulated by Roots (Polar Continental Shelf Project 1974, 12). Moreover, logistics and environmental considerations affected Arctic research to such a degree that, by the end of the 1960s, they had shattered Roots's faith in the achievement of experimental field methods:

> Aside from influencing the general seasonal nature of the work, logistics considerations tend to impress on a northern research party a schedule that is dictated not by the research itself, but by the problems of transportation, communication, and supply … The inflexibility of schedule often forces hasty work, with no chance for consultation or repeated observation; it leads to a habit of rushed, uncritical observation and description, of wholesale gathering of information with the hope that later analysis and reconsideration will sort out the irrelevant and find that nothing really relevant has been missed. It develops a "now or never" philosophy about the gathering of information, and creates a tolerance for incomplete, inaccurate or even shoddy research on the grounds that logistics constraints did not allow time or freedom to do a more careful job or to check observations and conclusions. (1969, 69)

As Roots put it to me in one of set of interviews, "The methodology of travelling determines what kind of information you get … A geologist working only with a helicopter might get a different picture of what the geology was than the one who was backpacking … So that's where research and methodology gets blurred. Sometimes the methodology determines what kind of results you get."[21]

Most telling of all, however, is the example from an attempt to construct a seismic refraction profile across the continental shelf by two PCSP seismologists, which resulted in a paper published in the *Canadian Journal of Earth Science* in 1971. For the authors, Berry and Barr, the "aim of the experiment was to obtain a crustal section from the coast of the northeast-southwest trending shoreline of the polar continental margin and slope to the water of the Canada Deep" (1971, 347). Unfortunately for Berry and Barr, this field experiment was hampered by bad weather and poor position control:

> In the event, the experiment had to be severely curtailed. *The authors appreciate that it is not normal practice to burden the reader with the difficulties of field practice but rather to concentrate exclusively on the results. However, the rigors of an arctic experiment and the uncommon problems encountered may be of interest to any other scientists contemplating a similar endeavor* … Suffice it to say that the unusually low temperatures encountered during April 1967, coupled with several unfortunate accidents, prevented our placing a recording station any further than approximately 220 km. from Houghton Head … Some problems were experienced in determining the positions of the shots and of the ice station, and as these problems will be a recurring difficulty in Arctic Ocean seismic work, they will be recounted in some detail. (348; my emphases)

The authors go on to discuss at length the difficulties of position control using a Decca navigation system on the ice to the west of Prince Patrick and Brock Islands (figure 2.4). Camp 200, an ice camp out on the ocean, was subject to an unexpected, radical path through April 1967 as various large-scale pressure systems built up in the ice (figure 2.5). It turns out that Roots had to make a special request to Computing Devices of Canada, who provided the Decca 6f Lambda

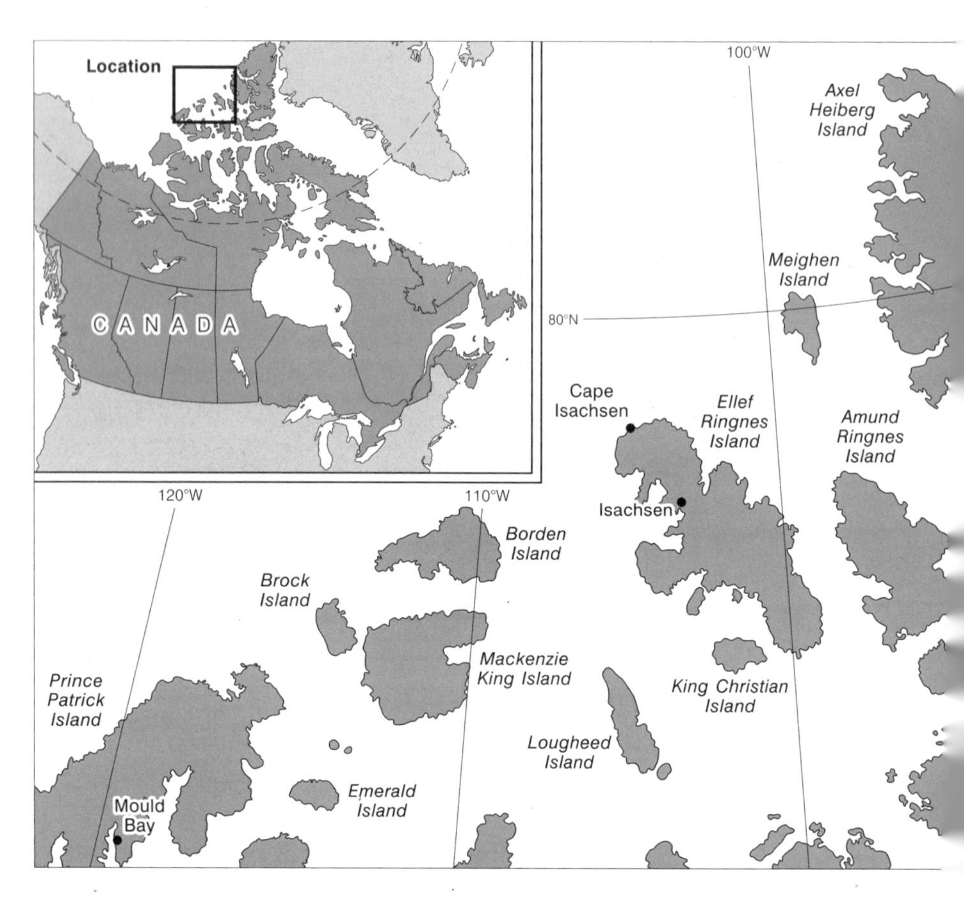

2.4 | Prince Patrick, Ellef Ringnes, and Brock Islands

chain-positioning technology to PCSP, for interpretation of the location of the seismic shots.[22] It appears that during the twenty-six-day period, Camp 200 moved twenty-five kilometres to the northeast along a "random walk" path (Berry and Barr 1971).

The accidents, as it happens, were precipitated by the destruction by fire of a shed at the Main Camp at Mould Bay, Prince Patrick Island. This resulted, a few days later, in a four-man helicopter party being marooned out on the ice for four days, following mechanical complications in flight, because as a result of the fire there were no operational aircraft to effect a rescue and no workshop space was available for aircraft maintenance. Bert Burry, a pilot on contract with PCSP,

2.5 | PCSP Ice Patrol, "Multi-year ice near Camp 200," 1970

thus described Mould Bay in 1967 as "a camp full of troubles" when providing a detailed account of the events after his return from the field to Uranium City, Saskatchewan.[23] The problem with such field projects is that the complicated logistics rely on the constant operability of the different components of the network. As Burry joked, it was fortunate that both Roots and the PCSP operations manager, Frank Hunt, were at the Mould Bay camp during this period of setbacks to the completion of the scientific program, as "it would be most difficult to believe all the foregoing unless one was there to witness it."[24]

But what is most interesting about this example is that it should have become evident that these activities never really comprised an experiment at all. By the strictures of the philosophy of experiment in the 1960s, it was, rather, a set of *structured observational measurements* to determine the seismic refraction profile that was being performed. However, this failure in experimental design became of less importance than coping with unprecedented environmental conditions. And it was the documenting of these circumstances that became a necessary, and *no less scientific*, activity after the field season had ended. It was in and through the quotidian enactment of field practice that the physical, epistemic, and emotional dimensions of experiment and observational measurement were *mangled*.

Rather less than deploying an authority of experiment, then, attempts by PCSP scientists to move to a conduct of Arctic field practice based on a model derived from physics merely re-introduced the authority of adventurous observation in different ways. The authority of field experiment was always precarious. Nor should we be surprised at this, as historians of science have shown that experiments form part of a rhetorical armoury like so many other scientific activities.

Conclusion

Despite their best efforts, those involved in the fieldwork of the PCSP had not discarded completely the epistemic baggage of the exploratory tradition and adventurous observation by the late 1960s. As to do so was deemed of great importance in order for the environmental sciences to become "scientific" in the sense of classical laboratory physics, this was a matter of some regret amongst many of those involved in the PCSP activities in the Arctic. Indeed, perhaps these pages should be read less as a narrative of the rigours of Arctic experiment, and more as an argument about the impossibility of the dream of field experiment, at least along the lines prescribed by the Popperian readings of earth and environmental scientists. However misguided such philosophies of field experiment may appear from the intellectual vantage point of early twenty-first-century sociology of science, I hope to have conveyed that the development and application of such practices were of great importance to the actors involved in constituting the PCSP.

It was a consequence of the multidisciplinary nature of these activities that scientists would be at the same field bases, on board the same aircraft, and even participate in the same field parties, as colleagues with quite different backgrounds and training. Geophysicists such as Berry and Barr, although undertaking investigations into seismic refraction, would often interact in the field with topographical surveyors or those constructing geomorphological maps. There would always be dialogue in the field between experimental practice and adventurous observation. However, this laudably diverse array of scientific accomplishment notwithstanding, the desire for field experiments permeated the institutional culture of the PCSP, and Berry and Barr as geophysicists enjoyed the consequent status within the scientific hierarchy of the base camps. Indeed, as the very founding purpose of the project was the investigation of the continental shelf, the emphasis here on this seismic refraction profile is particularly apposite.

This argument thus carries wider significance for historians of geography, as the debates about field experiment examined here anticipated some later debates about experimentation in physical geography, and especially in fluvial geomorphology and hydrology. Moreover, as I have argued elsewhere (Powell 2008a), it was crucial for the history of the geographical sciences in Canada that many of the PCSP field scientists had read undergraduate degrees in geography. It is therefore through consideration of such cases that a historical geography of the geographical sciences might be developed. Amongst the flourishing studies of geographical field cultures (Powell 2002; Lorimer 2003a, 2003b; Matless 2003), we must be wary of imposing a historiography that imposes too much uniformity in understandings of geographical field practice. Historical geographers of science have much to learn from investigations of particular cases of attempts to adopt experimental methods. Geographers of science might thereby be able to generate understandings of the geographical sciences that are more sensitive to the peculiarities of fieldwork.

As Ian Hacking puts it, "Experimentation has a life of its own" that often exceeds its bounds of initiation (1983, 150). As I have argued here, the engagement between geography and experiment has always been vivacious, not least because it depends on the contested question of "theory." This is not the locale to commence that history, but perhaps by accepting Trevor Barnes's (2001, 550) notion of "hermeneutic theorizing," which stresses creativity and catholicity in understanding, historical geographers of science might be able to re-appropriate a much-neglected slogan for our engagements with science studies that at least some field scientists were ready to adopt in 1969 – "By our theories you shall know us" (Harvey 1969, 486) – even if we no longer expect our theories to endure the various rigours of experimental control.

3 Base Cultures: The Spatial Organization of a Research Station

"What is this place?"
"It's called the Experimental Station."
That wasn't an answer.

Philip Pullman, 1995, *Northern Lights*, 242

Arriving in the Field

As many good ethnographers have noted, fieldwork is a personal experience that involves "finding our feet" (Geertz 1973, 13). Arriving at Resolute Airport in the High Arctic for the first time, it should have been unsurprising to find myself completely disorientated. I was not even sure whether this place, where I had disembarked, actually was Resolute. There had been no landing announcement on the First Air flight from Iqaluit. I knew that the flight was also to stop at Arctic Bay and, from my mental map, I had a feeling we might be there. Arctic Bay is closer to Iqaluit, but I discovered later that the usual flight itinerary is Ottawa–Iqaluit–Resolute–Arctic Bay–Iqaluit–Ottawa, so that the two High Arctic destinations are not repeated.

In any case, I must have looked bad, as a professor of geology from the University of Toronto, who had no connection to PCSP at all, asked me whether I was "in culture shock" and whether I had ever been to Canada before. The disorientation was compounded as, at this point, I had no idea of the procedures that were to be followed for arriving scientific parties. I knew from my historical work that the PCSP buildings were very close to the airport, but this was a new terminal and thus I could not be sure which of the nearby buildings formed the PCSP base.[1] It was actually a local individual, who turned out to be the mayor and owner of a hotel in the hamlet, who asked me what group I was from, and was able to indicate that there was a Polar Shelf representative at the airport.

When I met Gary, an employee of PCSP, he immediately asked me the obvious question: "Is this your first time in the north?" He then proceeded to tell me stories, which, in all likelihood, he had told many times before: "You don't need a seat belt, you're in the north. Hard to adapt, I know." After a few minutes of conversation, his jocularity seemed to modulate. When he realized that I was not the late arrival of some large scientific party, the majority of whose members were already in the field, Gary became more guarded. I was something else, much more disconcerting.

However, all this disorientation was absolutely necessary to prosecute the research, because it forced a reflexive moment about my presence in the field. John Law humorously relays some of the initial anxieties of the ethnographer, such as worrying about missing the action or being caught out walking purposefully in a circle from A to A whilst attempting to *appear* to be walking from A to B (Law 1994). In this case, I was conscious of the fact that I had to attempt to decipher quotidian activities, so as to be able to produce a narrative of scientific practices performed at PCSP.

Entering Fields

It was at this point, standing in Resolute airport for a few minutes, that I first began to comprehend one of the crucial factors that conditioned my fieldwork. I was alone and had no party to join up with. This was unique in scientific parties, as PCSP will allow only groups of *at the very least* two individuals to be put out at a field site. Moreover, a graduate student would *never* come to the field alone for the *first* time. A lead scientist will bring his or her group up to the base, and personally introduce them to the base managers, the base technical staff, and the cooks, so that, in the future, if graduate students or postdoctoral fellows have an itinerary different from the lead researcher's, they will know all the necessary individuals to communicate with in order to facilitate arrival at the field site. The fact that I was obviously a young researcher, yet I was arriving alone at Resolute for the first time, meant that from the very start my presence was viewed suspiciously, at least by the base staff. It was different for scientists, as they would often not realize that I was alone until I had chance to tell them or it came up indirectly through conversation about my project. In many ways, upon arrival at Resolute, I had to enter different *fields* created through the various communities that together compose the conduct of Arctic science.

My application for PCSP support had been processed through the standard scientific arrangements at the managerial level. The base staff had not been briefed about my presence before arrival. This meant that, until I was able to explain my intentions to them individually, initially just being around made them suspicious that I was either a young, undistinguished researcher unable to get anyone else to join my party, some sort of new surveillance mechanism sanctioned by management, such as a management consultant, or, worst of all, a British journalist. As Sarah, one of the base cooks, had it, "I don't like journalists because, like pilots, they think they are better than you." On one occasion, during a discussion between a research group and a base manager about whether a journalist should be allowed to visit a particular fly-camp, the scientist involved stated that he had "had the misfortune to sit next to her on the plane" on his flight up from Ottawa. Another, a Portuguese journalist, arrived unannounced at Resolute airport during my first field season and wanted to stay at the PCSP base. He was cold-shouldered by the scientists and PCSP staff, and ended up staying in the local *jail* for a few days (as a guest, rather than an inmate). This treatment was justified, according to older scientists, because they believed that far too many people come to the Arctic and "live off the good will" of the inhabitants. It was my impression that this was especially true of Resolute, precisely because it is a staging post for transport and communications in the High Arctic.

At the same time, though, I was on the list of incoming parties being offered support, and the fact that PCSP does not support journalists, writers, or other such "ne'er-do-wells" meant that, in principle, I had to be taken as a serious scientist.

Ethnographers commonly acknowledge that researchers are accepted more quickly by some members of a community than by others.

My acceptance by the scientific community at Resolute had a lot to do with two different Canadian scientists – Nicola, a senior government scientist, and Martin, a Canadian with significant Antarctic interests, who were both sympathetic to my project. Nicola, like many of the older scientists I interviewed, appeared to be very pleased that someone was researching PCSP from *outside* the politics of Canadian research, and my perceived "objectivity" was somehow reinforced by my being a British social scientist.

Some scientists would tend to be much more guarded, which was sometimes reflected in comments about my position as an outsider. There would often be jokes about my being British. An example would be a greeting of, "Well, you've brought the weather with you," whenever it was foggy, which was pretty much all the time. There was also, amongst some of the older scientists, a resentment of my being attached to the Scott Polar Research Institute (SPRI) in Cambridge. Many senior Canadian scientists appeared to associate SPRI with well-funded researchers from the natural sciences possessing all the latest equipment and instruments. With such scientists, and with senior base managers such as Andrew Walder, it was my MA from the University of British Columbia, and particularly, my attachment to "Ottawa U[niversity]" during 2002 that allowed for greater acceptance. As I will discuss, this perhaps reflected the widely held view in Resolute of PCSP as a Canadian *national* project.

In this chapter, I examine the organizational geographies of PCSP, that is, how PCSP is constructed through activities at different sites. It is then argued that attempts are made to govern the daily activities within these spaces through the construction of "rules" that are supposed to govern the conduct of those scientific practices that are supported by PCSP. These institutional rules attempt to manifest a culture of temperance in Resolute. As will be discussed, the accommodation of, and resistance to, the rules construct the practices of field science in the Canadian Arctic.

The Base Office: An Organizational Geography of a Communications Hub

The spatial organization of institutions matters in the embodied constitution of action (Garfinkel 1967). In order to understand the following

3.1 | Base office, PCSP Resolute, outside, August 2001

ethnographic material, the institutional arrangements of PCSP Resolute must be grasped from the outset.

My quotidian practices in Resolute generally involved spending time in the base office (figure 3.1). This forms the communications and logistics hub around which science in the Canadian Arctic is structured. The main office room is centred on a large desk, where the senior base manager sits. Behind the desk are the base radio and two telephones. In front of the desk are three chairs, where visitors are able to sit, as I did myself on occasion. Across the other side of the room is a computer terminal, where the more junior base manager is usually located (figure 3.2). A small room is encompassed within the main office, where a more powerful PC, with GIS capabilities, is situated. This terminal is used for constant updates of satellite photographs used for assessment of weather conditions. I would often observe the daily activities and conversations in the office by standing in the alcove of this small room.

3.2 | Interior of base office, PCSP Resolute, August 2001

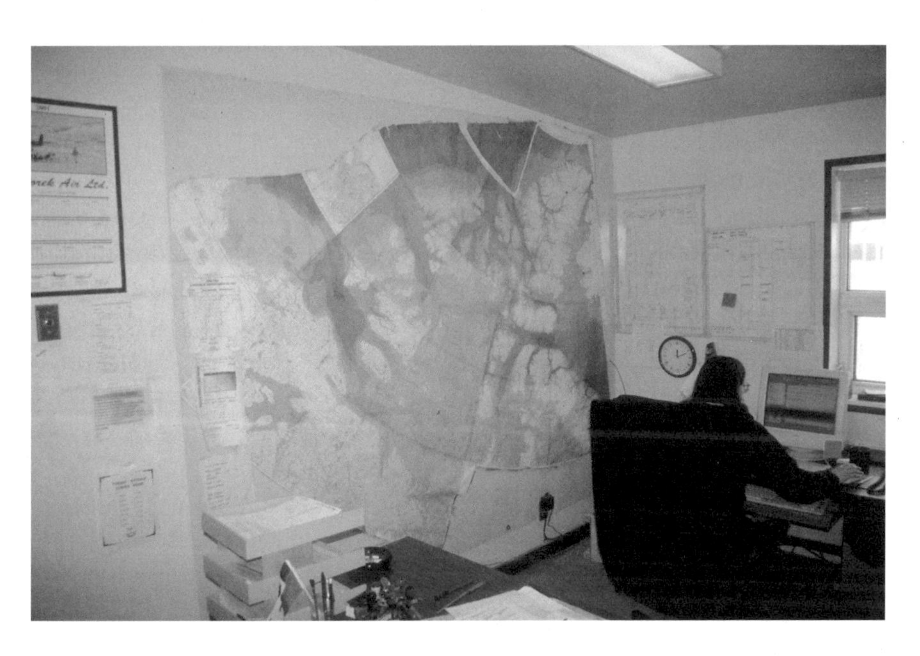

3.3 | High Arctic fly-camps, base office, PCSP Resolute, August 2001

In many ways, the office appears like that of any small-scale, corporate organization, with stationery, filing cabinets, a computer printer, and a fax machine. However, a number of accoutrements indicate the peculiar functions of this office. A large map is mounted on the wall, with red pins indicating the location of fly-camps (figure 3.3). This map is used for measuring approximate flying distances, with a long string attached to a black pin indicating the position of Resolute. The fly-camps are referred to either by geographical location, for example, "Bylot Island," or by the surname of the lead scientist. Fly-camps by surname are attributed only to senior scientists, and thus it is an informal mark of status to have a camp named in such a way. At PCSP Resolute, the scientists, as one of the base managers puts it, are known as "*beakers*, but it's not derogatory." According to a junior base manager, Yvonne, this is both because scientists use lots of instruments, such as measuring beakers, and also because of the character "Dr Beakers" on the Muppets television puppet show.

Although the PCSP maintains a base at Tuktoyaktuk, it has effectively been mothballed since 1998. This was justified by Andrew Walder, the most senior base manager, on the grounds of efficiency, as researchers in the western Arctic would now be able to fly straight into the field from Inuvik. For Andrew, this is much more financially viable than flying out to PCSP Tuk further north up the Mackenzie Delta first, and then being flown out by PCSP to the fly-camps. During the early years of the twenty-first century, many Canadian scientists and policy-makers argued that the scope of activities undertaken by PCSP had decreased dramatically in comparison with the early 1990s. Indeed, these debates continue to appear almost biannually (England, Dyke, and Henry 1998; England 2000, 2010). Regardless, as indicated on the huge map in the office in Resolute, the geographical expanse of the activities covered by the PCSP is still very impressive. During the 2002 field season, as well as the three helicopters and two Twin Otters based at Resolute, PCSP had chartered two helicopters to operate west of Repulse Bay (on Ellis Island), two helicopters for central Baffin Island, one helicopter at Cambridge Bay, and another Twin Otter in Iqaluit (figures 3.4 and 3.5).

There are two large diaries on the desk. These mark the scheduled arrivals and departures of, respectively, (1) *people*, staff, scientists and other visitors, and (2) *aircraft*, the Twin Otters and helicopters under

3.4 | Twin Otter, PCSP Resolute, August 2001

3.5 | Helicopter, PCSP Resolute, August 2001

contract to PCSP. There is also a folder on the desk containing lists of the "alpha projects," that is, those projects offered PCSP support, for the respective field season. Contained within the filing cabinets are more detailed files on each project. The two senior base managers, Andrew and Edward Freeman, index the files for each aircraft on contract in different ways. Edward does it alphabetically, and at one point he could not find the QNS file, for the helicopter with call sign "QNS," because Andrew had indexed them differently. He stated, "I'll have to change that!"

On the wall opposite the desk, and thus directly in the senior manager's line of sight, there are mailboxes for various fly-camps. This seems to allow the manager a quick way of checking which camps are out at a given moment, as mail from Canada Post for the respective parties is saved up. The mail is then collected by the scientists when they return from the field or, for those in the field for very long periods, it is taken out by the pilots when they visit a camp with new supplies or replacement faculty or students.

Another large room, adjacent to the main office, includes rarely used coffee-making facilities, a small photocopier, and a collection of 1:250,000 maps of the Arctic Archipelago. These maps are annotated by hand with the co-ordinates of old camps, information about the quality of possible landing strips, and notes about fuel caches.

During the day, there will be constant flow in and out of the office by pilots and scientists checking on flight plans, mechanics and base staff checking on duties, and perhaps local employees coming to collect wages. Throughout these quotidian activities, there is a constant hum of radio traffic in the background.

"Radio Skeds"

The co-ordination between PCSP Resolute and the fly-camps is organized around the twice-daily radio schedules or "skeds." These occur at 7 a.m. and 7 p.m. and involve each camp summarizing their condition when being called upon in turn by PCSP Resolute. The skeds are used to check that all is well at the camps, and perhaps to make requests for replacement equipment or for help with equipment repair. They are also used to check ground weather conditions when flying is planned in that particular area of the Arctic.

"XMH 26" is the call sign for PCSP Resolute's licensed frequencies, and "XMH 24" is the call sign for PCSP Tuk. The other radios, loaned out by PCSP to the researchers as they leave for the fly-camps, have portable call signs and thus are hailed by the fly-camp name, such as "Bylot Island." Weather conditions sometimes, even often, mean that fly-camps cannot communicate by radio with Resolute. The PCSP building at Eureka, Ellesmere Island, has a very powerful radio, and in such cases is thus used to relay messages to the "northern camps," on northern Ellesmere and Axel Heiberg Islands, when a base manager is there. Similarly, the staff at the Parks Canada base at Tanquary Fiord are used to relay messages to nearby camps when necessary.

At other times, "Two-Six," as in radio call sign "XMH 26," would transmit blind to tell camps that, if they could hear, they should telephone the base on satellite phones. These satellite phones, as of 2003, could finally be provided to every camp because of the problems with radio communication, and had been programmed with the main and emergency numbers of PCSP Resolute.

During my fieldwork, then, a number of new communication technologies were beginning to affect Arctic science. There were occasions, for example, when emails received at PCSP Resolute were read over the radio during sked to camps. The first time I observed this passing of information from an email to a camp at Alexandra Fiord, Ellesmere Island, Andrew Walder commented, "We've spawned a fucking monster."

In 2002, the morning sked was moved to 8 a.m. This was justified as giving the base managers more time to integrate the satellite weather report and the Resolute Airport Terminal forecasts that are also received daily at 7 a.m., as well as allowing the fly-camps more time to have their breakfasts and prepare a better account of the respective local weather conditions. This was especially important for the camps in the western time zone, such as Banks Island or Prince Patrick Island, where the schedule had been effectively at 6 a.m. When I asked Andrew about this, he said the change was "partly so people in the camps will actually get out of the tent to check the weather!" The pilots were never ready before 9 a.m. for flying anyway, he argued, so time is just wasted operationally with an earlier schedule. However, the downgrading of the centrality of the radio skeds is indicative of a changing field culture. The radio schedules are now less important than during earlier decades because the PCSP no longer collates all the weather information from all the fly-camps, as they did until the early 1990s.[2] Moreover, because of the constantly changing Arctic weather, and the introduction of satellite phones, weather information was simply being updated by all the camps between 7 a.m. and 8 a.m. anyway, thus making the skeds, for some at least, effectively redundant.

This change appeared to form part of a more general decline in the bodily discipline of the *enacting* of the radio schedule. Towards the end of the 2002 season, two base managers had a conversation about radio protocol. Edward argued that this "has gone to shit over the past couple of years." Yvonne thought this was because the radio reception had been so bad that particular summer, due to sun-spot activity affecting atmospheric conditions, that, consequently, so many messages had had to be relayed, that the protocols of confirming message reception, such as "I have you 5 by 5," had been all but forgotten. The solution, for Edward, was to transmit blind with a statement reminding all camps of the protocols. However, at the same time, I observed a few scientists maintaining exceptionally disciplined protocols that bordered upon the theatrical. These individuals tended to have high status within the base hierarchy, and I took this behaviour to be a concerted attempt to assert their legitimacy as the top *field* scientists.

Notwithstanding this occasional theatricality, as radio schedules have become less central, satellite telephones have thus revolutionized field practices. They have helped with safety considerations and emergency responses. In the past, if there was a problem with equipment or a polar bear sighting by a group in the field, PCSP Resolute would not know about it until the group got back to their campsite, where they would usually leave the radio. Walder now provides belt-packs for every satellite telephone, so that groups can take them out to their research sites from the fly-camp, whether by foot or ATV.[3] On one occasion, a group of two students at Sanagak Lake rang PCSP Resolute when their ATV broke down forty kilometres from the fly-camp. Although the signal kept cutting out, Walder was able to give them instructions to maintain the ATV, and thus "they were able to get the ATV twenty kilometres back towards camp before it finally packed in." I once asked Yvonne what happened, before the introduction of satellite telephones, when radio communications were disrupted by the solar flares. Her answer was that there were more camps then to relay messages, and planes could also fly over camps and transmit down to the ground if necessary, and that there was actually less solar flare activity in the past.

A second, perhaps unforeseen consequence of satellite telephones is that they are also now used for the more "private" aspects of conversations between fly-camps and Resolute, such as how many flying hours are left, or whether a team can afford to continue with certain projects and thus flights. I believe that this has helped reduce the collegiality of all the camps scattered across the High Arctic, as scientists are increasingly less aware of what is going on at other camps. This perhaps feeds into the complaints about increasing competition between groups that were noted by a number of older scientists.

Managing the Base and Its Rhythms

A scientific base in the Arctic requires specific logistical arrangements. The jet flights into Resolute airport structure the working week. The routings of these flights are Ottawa–Iqaluit–Resolute–Arctic Bay–Iqaluit–Ottawa, the "East Jet," on Wednesdays and Saturdays, and the "West Jet," Edmonton–Yellowknife–Resolute–Yellowknife–Edmonton on Saturdays. The jets are on the ground at Resolute only for around forty minutes for passengers to disembark and new passengers to board, with the engines left running because there is risk of freezing. An annual sealift for PCSP, and the other federal agencies in Resolute, arrives in late August. The sealift for the hamlet arrives a couple of weeks later that same month.

The quotidian rhythms of the base are structured by mealtimes, at 7 a.m., noon, and 5 p.m. At any given time, two managers are on duty in the office. The base managers work eleven hours per day, with shifts of 6:30 a.m. to 7:30 p.m. or 10:30 a.m. to 11:30 p.m., with two hours off for meals. The more senior manager usually takes the earlier shift. However, according to Yvonne, this can change if the jet is delayed, so that someone meets scientists as they come in from the south, and also if helicopters are delayed by weather, because they have to be continually "flight monitored." I also observed occasions when Twin Otters were flying late because they had been delayed by difficulties with the weather or with scientists, and a manager would still be on duty at 4:00 a.m., making at least sixteen-hour days occasionally necessary.

The job of base manager, as should already be evident, involves a lot of informal air traffic control and weather forecasting.[4] The ingenuity of the more experienced base managers in devising the most economical flying routes, and related task combinations, is truly astonishing. Although managers are supposed to have hour-long breaks for meals, they usually eat their food very quickly and then return to the office immediately afterwards. However, they do have frequent cigarette breaks just outside the office.

The weather in the Arctic is the biggest factor in modulating the decisions of base managers. As Yvonne argues, "You can plan as much as you want, but if it's foggy where you want to go and not where you are planning for the next day, then you have to change plans." Weather conditions are constantly monitored and assessed from satellite reports, Resolute Airport Terminal forecasts, and frequent, simple observations from the office windows.

Other technological developments have affected how PCSP deals with Arctic logistics. The introduction of global positioning systems (GPS) have reduced flying costs massively by allowing determination of precise locations of fuel caches and landing strips. Previous practice would involve pilots wasting precious flying hours attempting to locate such sites. Satellite photographs of weather have also had an important impact, because they have allowed pointless trips to be scrapped. Even during relatively recent years, pilots would still embark to locations and then have to abort when they reached areas with bad weather. Attempts are sometimes made to "pull" camps a day or so early if it is feared that the weather conditions would soon deteriorate in that particular area.

The weather can also restrict the ability of jet flights to land at Resolute. In turn, this affects the number of people on base on a given day and the resultant catering and accommodation arrangements. The malign impact is reinforced because, when the jet has not landed, the kitchen will not have been re-stocked. During July 2002, there was a particularly bad episode. The conditions at Resolute involved heavy snow and thick fog. A couple of hours before the jet was due, Yvonne made an announcement over the speaker system to the gathered scientists. First, "The West Jet has been delayed until about 3 a.m.," resulting in a twelve-hour delay. Second, and much worse, "The East Jet is no longer a jet and will be two 738s. These are due in tomorrow at 12:10

p.m. and 12:40 p.m. You will then overnight in Iqaluit, and will leave Iqaluit at 13:30 p.m. on Monday, to get into Ottawa at 16:15 p.m. You are on your own for hotels in Iqaluit and will have to book them yourselves." Andrew Walder was then audible over the intercom system, joking, "Thank you for flying First Air."

This caused much amusement and some anger among the waiting scientists, provoking much self-reflection about the practice of Arctic science. "Anyone who does Arctic science," joked Christopher Brown, a distinguished Canadian biologist, "deserves a medal." Brown had already incurred $7,000 in freight costs just to get his instruments up to Resolute, and First Air had lost some of his load. The financial implications of this delay were massive – his group of six would be charged $1200 for one night's accommodation in Iqaluit.[5] Moreover, because the East Jet was no longer a jet, there were problems for scientists in actually shipping their freight south, as the smaller aircraft have more stringent loading regulations and allowances.

In this case, the West Jet, supposedly due in at 3 a.m. failed to arrive, and Alan Merton, another accomplished Canadian earth scientist, went over to the airport during the night to check on the situation. The only person there was a Navigation Canada representative and he did not know whether the flight was airborne, so he telephoned NORAD to see if it had appeared on their radar. Merton believed that the delays have nothing to do with bad weather, and that the plane was being sent in the middle of the night because that would be the only time First Air could not use it elsewhere in the south. As these flights would continue to be delayed further each day, Andrew's situation reports would approximate the following form: "The latest lie is that WestJet is coming in at 2:30 a.m. tomorrow."

During my first field season in Resolute, PCSP managers were unsure how much fuel was at each cache and how well maintained various landing strips were. During 2002, Charles, another junior base manager, had completed a database that compiled much of this information. Each pilot was also given a digital camera to take photographs of the various landing strips and any buildings or facilities at each camp to be added to the database.[6] In addition, Charles had been out to all these sites, and had taken photographs and made inventories of the sites for the database. Andrew Walder explained, "We are trying to get all my, and Edward's, knowledge about landing strips and sites together and into a computer for the next generation. Maybe that's a good thing, maybe not. But the current generation [of base managers] don't have that." This database, with photographs of strips and various facilities proved its utility when a research group were moving from their traditional sites on Ellesmere Island to start a new project on Melville Island.

The base mangers also demonstrate good levels of other knowledge, experience, and general skills required around the base. Yvonne would often help out in the hangar by driving a forklift truck, quipping, "There are only two things that you need in this life: duct tape, for if it moves and it shouldn't; and WD40, for if it should move and it doesn't." Charles had been employed as a janitor in previous field seasons, responsible for the general upkeep of the base, before being promoted to a managerial position.

Managers come in and out of the field during a season, whether for vacation leave or to be transferred to other areas. The seasonality of work means that towards the end of a busy season, base managers might be occupied with obscure tasks that had been hitherto avoided, such as the moving of a pool table between buildings. In the closed season, senior managers like Andrew and Edward do planning work in Ottawa, such as considering applications and ordering fuel and supplies. Yvonne had some work in the fall of 2001 with PCSP, but not the next year, and Charles started a PhD in fall 2001.

Staff meetings were held during particular afternoons every couple of weeks, but as in most institutions, there was occasional friction between members of staff. An incident of tension between the managers, perhaps precipitated by my presence, occurred in the office regarding a request from a camp for a replacement Suzuki swing-arm ATV. After receiving this message, Charles began to arrange a substitute. Walder interrupted him, stating abruptly, "I haven't decided if they're going to get one. Just 'cos they ask don't mean they get. They broke two last year."

It became apparent that Walder believed that the researchers broke the ATVs by sitting two people on them. Charles disagreed that this could be the case because there were only two members left of that party in the field, replying, "I think you're mistaken." Amidst rising tension, especially with an ethnographer watching, Walder appeared anxious to demonstrate that *he* made the important decisions regarding the PCSP at Resolute.

3.6 | Hangar, PCSP Resolute, August 2001

The Hangar: A Space for Organizing Practice

One busy morning, during the July peak of the 2002 field season, Andrew Walder strolled into the PCSP office, remarking, "That's Polar Shelf: complete fucking chaos in the hangar. People going here and there." The PCSP hangar is the central space in the network of Arctic scientific practice (figure 3.6). This large aircraft hangar comprises an unpartitioned ground floor space, where pallets are stacked with equipment and supplies for fieldwork, ready to be loaded onto Twin Otter flights to the field sites. During the winter, when the base is closed down, this space houses all the ATVs, Skidoos, and other vehicles and equipment of the PCSP. Smaller rooms off the main floor space include a large workshop for PCSP motor mechanics, a much smaller workshop for helicopter and fixed-wing aircraft mechanics who are on contract (that is, they are contractually attached to specific aircraft), and the stores office. On the first floor, and around the sides of the

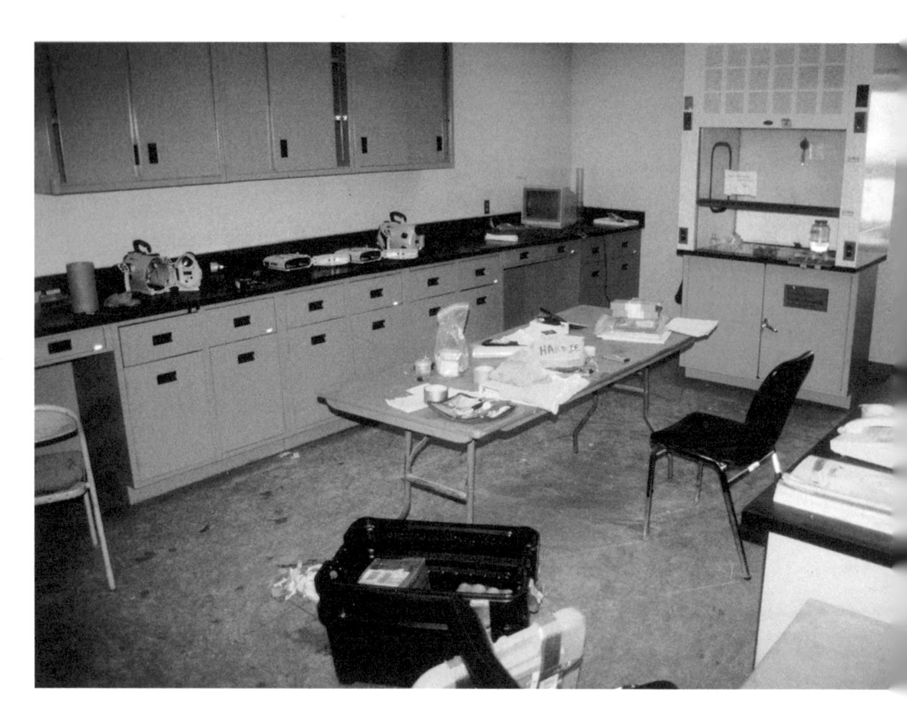

3.7 | Dry laboratory, PCSP Resolute, August 2001

hangar, are "cages" where scientists store equipment when not in the field, such as tents, water pumps, stoves, spare sleeping bags, and perhaps vacuum-packed food. ATVs or Skidoos, depending on the time of year and degree of snow cover, are parked just outside the mechanical workshop, so that they are easily accessible to scientists needing them to reach field sites on Cornwallis Island, or for loading onto a flight to a fly-camp. On this upper level, there are also dry laboratories for the scientists, as well as a computer room with two PCs (figure 3.7).

The stores office handles the shipping and receiving of supplies and equipment from the south. The front of the stores office contains a computer, and a radio with constant traffic. The shelves, towards the rear of the office, contain helmets, survival suits, satellite phones, over-boots, gloves, stoves, paper, and first-aid kits. Barry, a Newfoundlander, is the stores manager. In many ways this is the alternative hub of northern field science, as Barry co-ordinates the movements of assorted objects across the Arctic, as well as to the south and on across the globe.

When I asked Barry to describe his tasks, he replied that he has done twenty-eight seasons in the Arctic and had twenty bosses in that time. He summarizes his job as "babysitting" the scientists as they go into the field, by anticipating potential difficulties, because "90 per cent of beakers don't have a clue." According to Barry, the process runs as follows. A scientist sends Barry a list of equipment, and Barry changes it to "what they really need," such as when scientists forget to list a stove and propane gas. And when they return from the field, he performs essentially the same task, by arranging shipping tags and taking boxes over to the nearby First Air hangar to be shipped back to the departments and universities at which the scientists are based. Barry has a good contact in the central government stores office in Ottawa, and he can thus get replacement gear as necessary onto the next jet flight up to Resolute.

One of my conversations with Barry took place whilst he was dealing with a box bound for a biologist at L'Université Laval. Barry does the packaging personally and arranges the shipping, because the students usually do not know the procedures, and at the end of the season the professors have usually been out of the field and back in the south for long periods. Lead scientists tend to come up with the initial party, and then return home after a couple of weeks, leaving graduate and undergraduate students in the field. The scientist may return once during the summer, but the postdoctoral fellows or senior doctoral students appear to have substantial importance in the continuance of the program of research within the respective field parties. However, at the end of the season, the last members of a party to fly south, and who have thus been in the field for the longest, tend to be junior master's and undergraduate students. Barry thus has a crucial role, as the students tell him that boxes need to be shipped south, and he does the rest. After packaging, he takes loads over to the First Air cargo building, and charges the shipping costs to the respective scientist's account, and makes a double record on his computer database. As he stresses, this involves shipping boxes to scientists all over the world, such as Norway, Switzerland, and China. On this particular occasion, "The students did not even tell me they had things to ship, they just left these [boxes] in the hangar and I knew what to do."

As stores manager, Barry has ultimate responsibility for the different aspects of the base that are dealt with by the Newfoundlanders.

The Newfoundlanders are referred to almost universally on base, by managers, Inuit, and scientists, as "Newfies."[7] They also use "Newfie" to refer to each other. "Newfies" has become a term used on base to refer collectively to *all* mechanical and stores staff. Of the seasonal staff from Newfoundland, Barry deals with guns and radios, Gary with the building upkeep, Craig with the tents, and Roger with the ATVs and Skidoos. That Barry is critical to the functioning of Arctic logistics is demonstrated by a comment by Gavin, a climatologist: "Christopher Brown is coming in tomorrow [from Ottawa]. I only know that because Barry has moved his stuff from the back [of the hangar] to near the front. Barry knows everything."

The "Beaker" Building

Providing accommodation for scientists in Resolute is a main function of the PCSP. Whilst the pilots are accommodated in one of the original buildings in Resolute that, at one time, was the nursing station, the beakers reside in the main building. Constructed in the early 1990s, it includes a large number of twin rooms (figure 0.1). These are divided informally into male and female sides of the building, resulting from proximity to the respective bathrooms/showers. A main recreational room, with a TV, shuffleboard, and pool table, forms the focus to this building. When students came out from long residences in the field, with the fieldwork completed for the season, they would often cluster in front of the TV eating ice-cream and watching films as they waited for flights south. In the evenings, scientists might play a game of pool or occasionally watch TV. However, contact with the world *beyond* the field was relatively limited for scientists on base. Despite the fact that TV broadcasts might be the first opportunity to discover developments in the south when returning from isolated fly-camps, many scientists would *explicitly* attempt to avoid "hearing" any such news. This appeared to be another aspect of defining one's legitimacy to practise Arctic fieldwork.

There are few other opportunities for distraction for scientists. A small library adjoins the main recreational room. The holdings here contain copies of government publications, as well as various scientific publications, some MSc and PhD theses supported by PCSP, and a few trashy novels (figure 3.8). This would be used occasionally as a learning

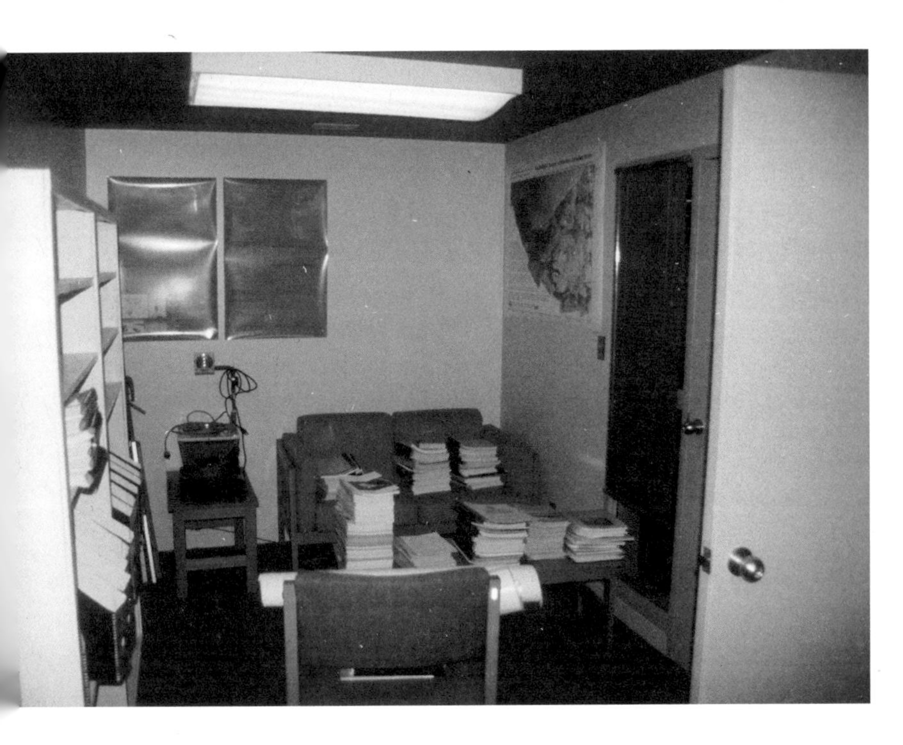

3.8 | Library, PCSP Resolute, August 2001

resource by scientists, and graduate students might peruse old theses. This space was also a useful venue for conducting semi-structured interviews with scientists.

The kitchen and dining area is adjacent to the recreational room, and the staff bedrooms and recreational rooms are located on the first floor above the kitchen. There are separate entrances and "boot rooms" for the staff and for the scientists and other visitors. An intercom system connects the whole base, including the offices and warehouse, and the main accommodation building, with a speaker in the lounge and dining areas. It is, however, impossible for scientists to hear messages if they are in the "beaker side" of the main building, where the bedrooms and bathrooms are located.

The kitchen staff have a crucial role in the reproduction of scientific research out of Resolute (figure 3.9). Like the Newfoundlanders, the two cooks, Sarah and Kristin, are on seasonal contracts. The cooks create a remarkable amount of excellent food, given the limited

3.9 | Kitchen and dining hall, PCSP Resolute, August 2001

availability of produce. Indeed, PCSP Resolute has a reputation for providing the best food in Nunavut. An older moniker for PCSP is actually "Polar trough." The "dessert tray" is so impressive that numerous scientists told me that it should get a mention in my book. A couple of scientists even complained that there was just too much food and that, despite the strenuous activity of field practice, they always put on weight during fieldwork.

The cooks get particularly angry when people do not show up at the beginning of mealtimes, because if someone gets a meal later it generally delays the washing-up, and thus extends their working hours. However, this is never expressed formally to anyone, and the mealtimes display a full hour, and thus it is usually scientists new to Resolute or scientists alone or in very small groups who commit this unwitting faux pas. This was also annoying at times, as the best opportunity to speak to scientists was over mealtimes, and the drive to collect the dishes could curtail many an interesting conversation.

My attempts to learn from mealtime conversations therefore would often be facilitated when there was bad weather, as snowstorms and fog prevented flying and scientists had more time to talk. Malign weather conditions were especially helpful just after or before a jet day, when the base would be full of frustrated scientists waiting to get into the field or back south. Senior scientists would notice me hanging around and occasionally taking notes, and most would pigeonhole me as a journalist before I had a chance to talk to them. The more self-conscious would offer a commentary on what they were doing if they thought I was watching. My presence at mealtimes, even without prompting, often resulted in conversations about the state of northern research and definitions of "science." The fact that I was working on the *history* of PCSP also unwittingly helped to reinforce many opinions that PCSP was winding down activities.

Conclusion

This chapter has detailed the spatial arrangements of PCSP Resolute, and has suggested that the organization of these institutional spaces influences the conduct of Arctic science. These spaces take on significance in structuring the interactions between scientists, logistical personnel, pilots, and other PCSP employees and visitors. Through this discussion, a number of emergent dimensions of the base cultures have been depicted. The particularity of the site of PCSP Resolute means that these cultures are created through a series of local and institutional rules, such as radio skeds and flight timetables. Sometimes formal, more often informal and unarticulated, these rules govern the constitution of Arctic scientific practice. In order to investigate this further, and how it affects individual actors, it is now necessary to look more closely at the composition of the rules that structure social life at PCSP Resolute.

4 Performing the Arctic Scientific Human

The differentiating terms we have used in the past were not designed to make visible the complex texture of knowledge as practiced in the deep social spaces of modern institutions. To bring out this texture, one needs to magnify the space of knowledge-in-action, rather than simply observe disciplines or specialities as organizing structures.

Karin Knorr Cetina, 1999, *Epistemic Cultures*, 2–3

"Horizontal Science": The Boundaries of the Polar Continental Shelf Project

Having presented the spatial organization of the base complex, it is now possible to begin to describe how these distinctive elements of field culture in Resolute affect scientific identity. In order to understand how cultures of nationhood, science, and temperance are performed within the PCSP, we need a better handle on who the scientists are in Resolute. How do individuals become PCSP scientists?

All scientists offered logistical support by PCSP have to subscribe to the ruling structures of the institution. These rules therefore mediate the relationships between the scientists, the PCSP management, and PCSP base staff at Resolute. In order to understand the organization of a large number of scientific projects in the High Arctic, it is necessary to comprehend the processes by which scientists are first offered support by the PCSP.

In November each year, PCSP receives applications from a vast array of clients with research plans for work across the Canadian Arctic. These are then divided into two main competitions, for government agencies and for universities. For the first competition, certain government departments are allocated so much funding each, usually in consultation with the respective "host" departments. The PCSP then allocates a priority rating to each application received, and these are then returned to the heads of the respective departments, who decide which projects should be undertaken. In these cases, therefore, PCSP provides a degree of "external" assessment for other government agencies deciding on which field projects to undertake. PCSP is also able to highlight which projects are more viable, in terms of location, because of a significant amount of activity in a particular area in a given season. However, as the costs are recovered by PCSP from the respective "host" departments, projects that have departmental funding will not usually be prevented from reaching the field by PCSP except for drastic logistical reasons.

The second competition, for Canadian universities and non-Canadian projects, is more rigorous and more controversial. There are obviously limited funds available for these projects, and unlike the government competition, they draw directly from the PCSP budget. The assessment is therefore different: a committee of professors who, according to base manager Edward, "excel in their fields," basically adjudicate on behalf of PCSP on which projects to support.[1] This is known as the Scientific Screening Committee (SSC), whose purpose is to assess the Form 11s submitted by prospective researchers. The logistics manager used to make these decisions, but the process has evolved, in recent years, to take greater account of peer review.

The SSC assesses projects "in terms of their science," as outlined in the proposal.[2] This is measured through researcher productivity (publications over the previous five years) and the training potential of a given project (measured by student involvement in the planned field season). Successful projects are then allocated different degrees of gratis support, depending on the nature of the proposal, such as meals and accommodation at PCSP Resolute, the loan of ATVs, tents, stoves, and other equipment for the field, and most critically of all, "flying hours" by helicopter or Twin Otter. A project is granted so many "hours," and if they are exceeded, perhaps as the result of poor planning but almost always because of bad weather (e.g., fog prevents a pilot landing at a site, and thus time is spent circling or aborting a flight), the excess cost is recovered by PCSP. It is also possible to be offered a combination of gratis support and cost-recovery support. According to both Edward and Andrew, base managers become involved in decisions only if they think that a project simply cannot proceed for strictly logistical reasons, such as a helicopter or Twin Otter cannot be made available at a respective field site on given dates.

Within the SSC assessment, the PCSP Arctic-Antarctic Exchange Program allows preferential consideration for projects that involve international exchange with researchers who normally work on Antarctica. The idea is that PCSP provides free support to an Antarctic researcher to collaborate with a Canadian party, and then the Canadian lead researcher will enjoy reciprocal support in Antarctica. The SSC also considers applications by non-Canadian researchers and by "private interests," but such projects are always charged at the full cost-recovery rate. As a non-Canadian researcher, my application thus went through this process twice, for fieldwork in the 2001 and the 2002 seasons. Examples of private interests supported by PCSP include journalists, documentary film crews, or scientists attached to oil companies.

There is a third, much smaller fund for university researchers who wish to include traditional ecological knowledge (TEK) in research plans. As well as sole TEK projects, it is therefore also possible for a researcher to have a successful main bid alongside a TEK project. These TEK projects tend to be for research in the physical sciences that incorporates Inuit knowledges in some way, and thus does not usually cover what would be construed as "social scientific" studies of Inuit societies. It is worth noting that the successful Canadians under the TEK competition have also tended to be successful in the PCSP Arctic-Antarctic Exchange Program and main competition as well. There is therefore a degree of cynicism amongst scientists about the *perceived* diversion of limited funds away from more traditional scientific research in the High Arctic by Canadians through these new programs. However, perhaps as the result of the politics of relations between Inuit and scientists, as will be discussed further in chapter 5, this resentment is articulated more strongly against the PCSP Arctic-Antarctic Exchange Program.

Given that such decisions influence individuals' research careers, as well as being central to the purpose of the PCSP, questions about the methods of PCSP decision-making were ethnographically problematic. When I asked Edward casually about how project support operated, he referred me to the director. It became apparent during fieldwork that this is a matter of some concern to many scientists, and it is very difficult to discover under whose authority such decisions are made. It is therefore useful to draw from comments made by Heidi, the PCSP director, when I talked with her about this at PCSP Resolute:

> There are lots of rumours about us selling out to big $40-million projects. But Polar Shelf is a not-for-profit organization. We only support meritorious science. I will not compromise the reputation of Polar Shelf by supporting crap science. Of course, there are problems with any committee in terms of vested interests. But that is the same with NSERC [Natural Sciences and Engineering Research Council of Canada]. The Canadian research community is very small. The Canadian northern research community is very, very small. We can never have complete objectivity on the Scientific Screening Committee. Whoever is on there has their own personal baggage. Members of the committee leave the room when their own research is being discussed, but, even so, they work in teams, so they want project X supported, or they want to share project Y's data. But the people on the committee are at least aware of this situation. And the committee has been very helpful to me in making decisions about who to support. The committee still works in the same way as in 1990. All projects are ranked. [Grade] "One," offer support. [Grade] "Two," offer some support and some cost-recovery. Or [Grade] "Three," do not support, usually because there are problems with the methodology or science, or there is no student involvement, or there are no publications in the last five years. We have also awarded "Threes" to applications under the Traditional Ecological Knowledge Program which do not fit the criteria. Although, sometimes "Ones" cannot be supported for logistical reasons, such as we aren't going to be in that area

this year, or we can't get a helicopter available when you want it. There have been some appeals at decisions. Some very strong ones. Especially when we get smaller budgets and there is not enough money to go around. But it is a *myth* that the Scientific Screening Committee decides what *sort* of support should be offered. They just rank the projects based on their science. There are usually two representatives from the Scientific Screening Committee who also sit on the Polar Shelf Advisory Board. Polar Shelf pays the expenses for the Scientific Screening Committee to come to Ottawa, and usually the meeting is on the Monday, and the Advisory Board on the Tuesday. Scientists are often willing to make a huge time commitment to the organization. We have been lucky with the commitment of the community to the organization.[3]

It should be obvious, then, that there is tension here in attempting to adjudicate on a definition of "meritorious science." It would also appear that there is an implicit ranking of projects within the "Grade One" category, in that certain projects are never "outside" the PCSP area of operations, whereas other projects are, and yet both are adjudicated as worthy of support.

This raises a number of other interesting points. First, there is no recompense for PCSP if a project cancels unexpectedly, such as because it has not made a successful bid for Research Council funding. This can, according to Edward, leave PCSP with unwanted aircraft, but it is usually not a large problem if a small project cancels. Second, notwithstanding official PCSP policy, as non-Canadian projects would be charged full cost recovery, base managers would take decisions to try to keep costs down for Canadian university scientists. A group of American scientists were fogged in at Beechey Island on one occasion, but Edward stated to Yvonne, "They're paying, so we will try again to get them." On another occasion, David Smith, a Canadian bio-geographer, was trying to get out from the field, but so were a documentary film crew from National Geographic. Both were picked up by the same flight, so that, Edward stated, "we can use them to help pay for David's flight." This slight disadvantage of non-Canadian projects applied to me as well, at least in 2001, so that when I asked Edward whether I would be able to get on any more flights, he replied,

"No, they're full, but I really should charge you a percentage for those [previous] flights anyway, as you're a foreigner."

Erecting Boundaries around Arctic Field Science

I leave these matters of PCSP policy to make a number of observations about how PCSP defines "science" and the consequences of this for Arctic fieldwork. Drawing on sociologist Thomas Gieryn's notion of "boundary-work" (1983, 1999; Nader 1996), I argue that many important struggles about what constitutes worthy scientific practice are enacted at PCSP Resolute. In an example of both boundary-work and the politics of Canadian northern research, Quentin Lefebvre relayed a story of how a colleague was refused support by PCSP.[4] Steven Tyler wanted to get a flight to Grise Fiord, but his application was unsuccessful. The director allegedly told him that PCSP does not support social science.[5] He then offered to pay cost-recovery, but got no response.[6] And yet here I was in Resolute apparently conducting "social science." How could this be so? This indicated an inconsistency for Lefebvre in what sorts of activities PCSP decides to support. But how, he wondered, was it possible for scientists to raise this issue?

Another example was when, during the 2002 season, a sign by a junior base manager appeared in the computer room, requesting that large downloads be prohibited until after 8 p.m., as well as stating, "Please give priority to work related activities (especially science)."[7] Whilst I read this primarily as a warning to those undergraduates and pilots who used the computers to listen to online music broadcasts from Toronto,[8] I also saw a barbed piece of boundary-work as positioning my work as "not science."[9]

However, the most important incidence of boundary-work was the reaction of Andrew Walder to mention of the author and adventurer Jerry Kobalenko. This episode shows how PCSP base managers use the rules of PCSP support to erect boundaries between "the scientist" and "the amateur." It is critical for current base managers, like those earlier generations of PCSP scientists discussed in chapters 1 and 2, to repress the "adventurous" aspects of Arctic fieldwork to maintain their sense of legitimacy.

Kobalenko published a book about his expeditions to Ellesmere Island in 2002, entitled *The Horizontal Everest*, just before I returned to Resolute. It should be apparent from the title that Kobalenko is attempting to transfer codes of heroism and masculinity from the Himalayas to the Arctic. The book includes a serious slight of PCSP:

> Resolute is actually two villages: an "unlovely huddle" of interconnected government buildings near the airport, and the Inuit village. The two are connected by a four-mile gravel road of excruciating dullness that the truly desperate sometimes hike to kill time … Today it is sometimes possible to hitchhike back from Ellesmere, but planes to the island are usually full, and to show up in Resolute without firm plans is to risk never getting out … But sometimes all options fall through. You sit in Resolute for a week, camping behind the airport or watching your dollars fly away at one of the local hotels. The charms of Resolute wear thin. You have worn thin on Resolute's citizens. The airline managers stop looking up when you walk into their office for news. *The Polar Continental Shelf Project, with its many Ellesmere charters, will not accept your offers of good coin, leading you to be somewhat cynical about that science agency's well-publicized financial straits.* (Kobalenko 2002, 4–5; my emphases)

Having noticed this comment on PCSP, I found a moment to ask Walder about it. His visceral reaction indicated how seriously boundary-work is undertaken by PCSP.

In the first place, Kobalenko refuses to adhere to the rules of northern activity as established by PCSP. According to Walder, he is not a field scientist and is therefore just an adventurer who must pay like every other tourist. As Walder put it, "He pissed off First Air, then Canadian [Air], and then us … He reckons he should be able to pay $100 to get on a First Air flight rather than pay the $6,000. He pissed off all the hotels in Resolute as well. They won't let him stay anymore … he just says, 'But I'll give you great publicity in my book.' Once they caught him sleeping in the First Air hangar, then he was camping out in the field."[10]

Second, Kobalenko attempts to draw epistemic authority from the supposedly harsh environmental conditions of Ellesmere Island during the summer: "He thinks he's a 'great man,' but he isn't, he just goes chasing cairns. He wouldn't last a week up here in the winter. Russell Merris is an old man and he walks around Ellesmere on his own for six weeks every year. What did he call that book? Yeah, it's a *real* 'Horizontal Everest'!"[11]

But most important of all, Kobalenko failed to live up to his promises of accomplishing a program of research. After applying to PCSP for an investigation of the field sites of various nineteenth-century explorers, Kobalenko allegedly failed to undertake the program for which he had won logistical support.[12] This led Walder to exclaim, "*Polar Shelf don't do tourists!* There are two charter businesses in town. He doesn't do any research. He doesn't really know where the [archaeological] sites are, so he will come and try and get it out of us. But I won't tell him. He never does what he says. I think we [PCSP] supported him one year because he said he would do some 'science.' But he didn't do it, so we never flew him out again."[13]

Note that in this context, "science" for Walder, and thus PCSP, would include properly prosecuted *historical* or *archaeological* research. Interestingly, because there are now flight charters running to the only two communities in the Canadian High Arctic, Resolute and Grise Fiord, PCSP will no longer fly researchers to these sites. The unexpected result is that social scientists have been decreasingly able to conduct research in the High Arctic, because SSHRC will not cover such research costs, and there is thus no way to replace the essential logistical subsidy provided by PCSP. This has left some social scientists claiming that they have heard unofficial policy statements from PCSP that they do not support social science.

After this conversation, Walder joked that he would write his own book entitled *Horizontal Science*. When I saw him around the base thereafter, he said things such as, "I hope you're taking notes for my book! It'll be a bestseller. I will go on Larry King to talk about it. Yvonne can do Oprah to do the other side of the scoop!" This was generally very amusing. A number of chapter titles were established, such as "Chapter 3: Tales of the Mess Hall," "Chapter 5: Tales of Tanquary and Truro," and "Chapter 8: Isachsen." A number of Walder's "words of wisdom" were also to be recorded, such as "You don't get anything by cheating in life." On another occasion, Andrew's "tongue was coarse from licking lichen." However, an unforeseen consequence of these jokes was that Andrew remarked in a crowded dining hall during dinner, "Where's Richard Powell? He's recording my diaries." This drew attention from *everyone* at the base at this point, and appeared to anger some other staff and scientists, as if Walder was trying to monopolize my study.

Another example of the boundary-work undertaken by the management of PCSP is the relationship with NASA. Edward discussed how PCSP were happy to support NASA when they first came up to do the Mars Analogue research at Devon Crater, Devon Island. Taking as its premise that Devon Crater is the environment on earth that most closely resembles Mars, this NASA project focuses upon forms of life that exist in the rocks and freshwater of the High Arctic and thus might provide insights for the search for extraterrestrial life on Mars. The activity on Devon Island is mainly American, although some Canadian university professors and the Canadian Space Agency are collaborating with NASA.

However, when the Mars Society, described by Edward as "a left-wing organization of crackpots,"[14] became involved, PCSP "attempted to separate the science from the Mars Society."[15] More charitably, the Mars Society is an American philanthropic organization that wants to send the first people to Mars to set up a frontier colony. They thus spend most summers at the Devon Crater building structures in "spacesuits." But in consequence, during the 2001 season, PCSP support was limited to three flights and the loan of some ATVs, with the bulk of the NASA flying contract being filled by Bradley's, a private contractor.[16] Edward believed, during summer 2001, that NASA might try to gain PCSP support again in 2002 because Bradley's were planning to pull out of Resolute, but by the following season their withdrawal failed to materialize.

The behaviour of some NASA scientists can be illustrated by an individual in Richard Baker's group. Although Baker, a distinguished Canadian environmental scientist, has significant ties to PCSP, within ten minutes of disembarking from the Ottawa flight in Resolute for the first time, his colleague from NASA was in the PCSP computer lab, using one of the two PCs available to scientists, and the PCSP Ethernet cable connection with his own laptop, as well as loudly monopolizing the only telephone access point for scientists. He was thus not only making minimal attempt to interact with other scientists or staff at PCSP, he was visibly impeding anyone else attempting to make use of the PCSP resources. And he was able to stay at PCSP in the first place only because Baker was a Canadian.

The Social Hierarchy of Scientists on Base

During my fieldwork, some polar researchers, perhaps more used to field science with organizations such as the British Antarctic Survey (BAS), would comment on the more traditional, even "amateur" approach to field science still detectable within the practices of the PCSP.[17] As I have argued, the rules regarding PCSP support attempt to police such "amateur" or "adventurous" tendencies amongst contemporary field scientists. At Resolute, many senior scientists, such as Richard Baker,[18] argued that this older, amateur type of science had been purged from PCSP. I was unsure that this was completely so. I observed a number of groups, for example, who were very excited to be going out "camping" in the field. Such individuals were associated with both government departments and universities.

Perhaps emerging from these distinctions about proper field conduct, scientists certainly appeared to be dealt with through an implicit hierarchy. It may have operated subconsciously, but it was an observable feature of social interactions at the base. The older, male university professors were, assuming *ceteris paribus* in weather conditions at the respective field destinations, given priority for flights out to the field. Moreover, such scientists enjoyed other subtle privileges. Even prior to arrival, distinguished scientists, such as the holder of the only NSERC Northern Research Chair to work in the High Arctic, were spoken of with great deference. All the pallets with field equipment for the Northern Research Chair were unusually laid out in the hangar *two* days before he was even due to arrive. When weather conditions at his research site meant that PCSP wanted his group out first thing in the morning, he was still able to delay until his favoured afternoon departure because he had not finished repacking his loads. Less respected scientists would be forced to adhere much more closely to the timetable desired by PCSP.

The treatment of distinguished scientists in this way could cause disillusionment among others at the base. When an eminent Canadian scientist arrived at Resolute, for example, and was being feted with his large group of students by the cooks in the kitchen, it was obvious the effect this had on other scientists who believed themselves to be less accomplished (or thought that they were perceived as such by their peers). Such scientists would always begin, on such occasions, to talk about their spouses or returning home.

But above even successful Canadian university scientists in aircraft priority were foreign scientists, such as groups of Japanese, Scandinavian, or American researchers, likely because they were paying for services at the non-Canadian cost-recovery rates. However, such foreign scientists, unlike the senior Canadian university professors, were not generally as highly respected by other PCSP clients, and would often be left to their own table during meals to converse amongst themselves.

In the next rank in this informal hierarchy came researchers from federal departments that PCSP deemed important, such as the Geological Survey of Canada or the Canadian Remote Sensing Centre. Then, at the bottom, came an eclectic collection of small-group researchers from the less prestigious Canadian universities and the Canadian national museums.

Such hierarchies may reflect trends in the organization of any scientific activity, especially as neo-liberal competition began to influence the practice of scientists in recent decades. But these social practices are very different from PCSP in the 1970s and 1980s when, according to older scientists, there was less distinction among the PCSP clients, and more of a perpetuated division between those who enjoyed continuous PCSP support for many years, and those who would receive none, perhaps because they had angered a previous director at some point.

Pilots and the Ability to Be "Outside" the Rules

Unlike the employees of PCSP and the scientists depending on PCSP support, the pilots tend to enjoy relative freedom from the rules structure of the base. PCSP depends on the pilots of Twin Otter fixed-wing aircraft and helicopters to make difficult landings in hostile flying conditions that are critical to Arctic field science. Kenn Borek, a company based in Calgary, Alberta, held the PCSP flying contracts during both of my field seasons. The pilots usually work a three-week rotation and then return south. The majority of the Kenn Borek pilots were in their twenties or early thirties, and displayed the sorts of codes of competitive masculinity that Linda McDowell records amongst merchant bankers in the City of London (1997). Frank, for example, a pilot from the Yukon, attempted to display as little empathy as possible when he heard that a female scientist had been hurt at a fly-camp in an ATV accident: "Oh, that chick got hurt? That's what happens when you've got your head in your ass."

Moreover, this sort of macho insensitivity was surpassed in displays of contempt for PCSP regulations, because of perceived autonomy that came from being formally employed outside the institution. For Frank, orders came from Borek, not PCSP. When asked by a base manager one evening for his flight schedules and load records for that particular day, Frank replied, "I don't do Polar Shelf paperwork because it's not in my job description. I work for Kenn Borek, not Polar Shelf. Why should we respect you? Because you are our elder?" Finally, on a flight back from Cresswell Bay, Somerset Island, to Resolute, Frank took it upon himself to provide some in-flight entertainment for his passengers by performing some aerial acrobatics. An undergraduate field assistant was very upset by this, and an official complaint was made, and many other female scientists thought this behaviour was unsafe and unprofessional. Other pilots began to refer to this as "the infamous Cresswell Bay flight" and were worried that if news got back to head office, Borek might lose the PCSP contract.

However, not all pilots displayed such extreme behaviour. Another pilot, Ben, was exceptionally hard-working and was consequently very popular with managers. On one occasion, after flying late the night before, he was due under terms of contract to sleep until 9 a.m. and then to report for duty at 10 a.m. He showed up for duty at 8:05 a.m. It was interesting that Ben was one of only two older pilots and had a family with small children in the south.

When I asked Edward about the pilots, he argued that in the early years of PCSP, pilots had had to take risks. However, an experienced pool of pilots had been built up, and the embodied knowledge in this collective was crucial for PCSP to complete its tasks effectively. Professional and experienced pilots are, for Edward, critical in the completion of successful field seasons by the PCSP.

During 2001, the pilots were very well integrated into the social life of the base. The pilots and aircraft mechanics on contract were accommodated in the "old building," which was the original PCSP dormitory during the 1970s. The pilots had meals in the main PCSP kitchen.

Towards the end of the season, the remaining pilots and mechanics were moved over to the main beaker accommodation building, and the old building was closed for the winter. As pilots and mechanics were no longer required on contract, they were moved either to the Narwhal Hotel across from the PCSP base or to the South Camp Inn in the village.[19]

Social Dynamics at PCSP Resolute

In June 2002, there were a number of rule changes on base. First, it was decided to accommodate all pilots and mechanics at the Narwhal Hotel. This further lack of presence made the base much quieter than in my first season. Second, no one was allowed to help out in the kitchen. And, third, scientists and pilots were not allowed to go upstairs in the beaker building on occasional evenings to visit the PCSP staff common room.

One senior scientist's reaction to these new rules, after he arrived in the field in 2002, was hilarious. He listened intently as a graduate student explained them to a group of scientists over lunch. But when another student asked why a particular change was introduced, the senior scientist exclaimed, *"Don't ask why!"*[20] Moreover, the scientist was very surprised to hear that the pilots were staying at the Narwhal Hotel. The conversation at this point was illustrative:[21]

GRADUATE STUDENT: It costs $3,000 per day [for the pilots to stay at the Narwhal Hotel], apparently.

SENIOR SCIENTIST: Who pays for that?

GRADUATE STUDENT and SCIENTIST B (*simultaneously*): Polar Shelf!

SENIOR SCIENTIST (*very surprised*): Why?

SCIENTIST B: Because there is a leak.

SENIOR SCIENTIST: Bloody hell! That building has leaked for twenty years. Five years after it was built, it leaked.

SCIENTIST B: Obviously it's just an excuse. The pilots hate it, because everyone smokes over there. And the cook at Narwhal isn't very good. It also makes it really difficult to find anyone over there now. Lots of the pilots don't want to fly for Polar Shelf anymore.

GRADUATE STUDENT: The pilots are *now* allowed to come over here to make their lunches. Narwhal used to give them frozen submarine sandwiches, which thawed out and then got all soggy.

...

SCIENTIST B: It was much better with the pilots around, because there were more people to chat to. Although the same stories got more amazing every year.

Although it was difficult to discover the exact reasoning for these decisions, my sense from other informal conversations was that this was a conscious attempt by management to reinforce the *temperance culture* of the base, by reducing opportunities for dissent through the interaction of the pilots with student scientists and staff. Certainly, in 2002, the planning and convening of drinks parties by pilots and some scientists became more ingenious.

Polar Bear Stories

Having understood how PCSP attempts to regulate field science in the Arctic within the architectural spaces of the base, it is now important to discuss observed activities of scientists at the field sites. Polar bear stories often dominated conversations over dinner, especially in 2001, perhaps because a number of different camps across the Arctic had reported multiple polar bear sightings. An unusually early ice thaw that summer had resulted in changing behaviour by the polar bears.

Talking with one research group over lunch, I got a real sense of the fear of some graduate students about polar bears. One student thought their supervisor was neglectful for not giving more advice and rifles for dealing with bears. As he argued, "It is hard to hit a polar bear [with a rifle]. If you just hit them on the arm, you just have an angry bear! With a rifle, a scope, and tripod, which are all heavy, plus equipment, waders, and daypack, it is very difficult to carry everything, especially when you have to walk over nine kilometres of hummocks to get to your field sites." If they are inexperienced, he continued, most people are given shotguns rather than rifles, but he would prefer a pistol, because it would be easier to carry. But it is a moot point whether a pistol would be enough to stop a polar bear.

However, in a strange way, polar bears provided my first access to the observation of field scientists in action at a research site.[22] Nicola, and her postdoctoral assistant, Elizabeth, were in need of "a rifleman." The two women were botanists developing an interactive guide to enable visitors to the Arctic to identify plant species. The problem for Nicola was that there are field protocols for dealing with polar bears safely, involving constant observation from high ground, but, as she put it, "When trying to botanize, looking up is a bit of a bore." So, during my very first breakfast at Resolute,[23] I was recruited as a "bear man." As soon as Elizabeth was out of earshot, Nicola told me, "I'm not bothered by bears at all, but Elizabeth is, so you're very useful. Thank you." Nicola, as the more senior scientist, appeared to want to disassociate herself from the stigma of being afraid of polar bears, as it might undermine her legitimacy as an Arctic field scientist. There may also have been a gendered dimension to this affirmation. I later found out that, earlier in the field season, they had encountered a polar bear near Churchill and had had to wait in their vehicle until it left, and that they had already been rescued, on another occasion, from a bear by two Inuit hunters.

Nicola and Elizabeth had decided to take a vehicle to explore the area around Resolute, because, although they had been allocated five hours of helicopter flying, they did not believe the aircraft would be available to them "any time soon." Again, this reflected their perceived low status in the hierarchy. Other projects were deemed more important than this sort of botanical fieldwork, so those researchers would be higher up the "pecking order" when the helicopter did become available.

As a result, Nicola collected the keys from the base office for a "green vehicle," and we initially got into a green van. On finding that the keys were already in the ignition of this van, Nicola instructed us all to get out. We then located the green PCSP Chevrolet Suburban that the team had been allocated. Nicola drove eight kilometres to the hamlet of Resolute, where the botanists decided "to botanize." It was very foggy and raining slightly. The procedure at the field site involved Nicola and Elizabeth walking around the area, and then deciding upon their respective tasks. They then split up. Both lay on the ground to identify particular plant specimens (figure 4.1) by taking photographs and samples for analysis back at the laboratory at the base. The entire visit at the field site took around thirty minutes.

4.1 | Botanists in the field, near Resolute, August 2001

Afterwards they announced that this visit to the High Arctic had been worthwhile, because they had already found interesting data on this first trip out of the base. Nicola noted, "This is always the way: you get lots when you don't expect to." She called this process "creative botanizing." We then moved on to the "Thule site," near Resolute Lake, which has been restored by archaeologist Robert McGhee as a National Heritage site for tourists (McGhee 2001).[24] It was raining heavily at this point. Nicola fetched tools whilst Elizabeth did most of the "botanizing." When sampling at this location, Elizabeth got much closer to the plants in order to be able to distinguish certain types from others. One grass sample that Elizabeth collected was apparently important, because it will settle a debate about the species when the DNA is tested.

There was then a discussion between the botanists about whether to continue to another site on the other side of the island. The conversation revolved around the weather, they would be "wet and cold" if they continued, but they also factored in the cost of the vehicle hire at hourly rates, and it might also be impossible to get access to the vehicle again. Nicola's attitude to Arctic field science involved a high degree of humour. As she remarked, "When I was younger, there always seemed to be less vehicles available. I think they see me as a wise old woman now. Either that or they have got more vehicles." On another occasion, as the weather began to deteriorate rapidly, Nicola joked, "I'm a civil servant, so we will be back in time for lunch." Whilst in the field, Nicola often asked Elizabeth, as a surrogate it appeared, whether she was too cold or wet and if she wanted to return to the base.

Both botanists, along with others I met at Resolute, bemoaned the fact that the field expertise they deployed is no longer sufficiently present in the scientific community. Having been trained at the molecular level, they argued, biological scientists now no longer know enough about taxonomies.

When we got back to PCSP, we went to their "dry lab," in Room 202. Each scientific party usually has access to one. The allocation seems to be by seniority of scientist and by size of party, so that a "star" university researcher with large group of students gets one of the bigger labs. At busy times, there was a lot of competition for them, whilst at quieter times a small group of two or three individuals could be making use of three labs.

Nicola and Elizabeth arranged the plant samples on a laboratory surface, together with wax paper marked for places such as Oslo, Churchill, and Ottawa. Around the lab were tools, sample bags of plant specimens, their field daypacks, and a suitcase with equipment – although a second suitcase had been lost by the airline on their flight into Resolute.[25] Elizabeth had a laptop computer, which was used for data entry, and the name and location of the plant specimen was recorded together with a grid reference and a textual description.

The ingenuity required to conduct such fieldwork is striking. Data storage for transportation requires polystyrene cups, wax paper, and plastic sandwich bags to be borrowed from the PCSP cooks. According to Nicola, fieldwork out of Resolute often involves camaraderie, multitasking, and "doing things through the back door," although she felt it

4.2 | Hydrologist performing measurements in the field, near Resolute, June 2002

was always important to "keep the office updated with where you are going for safety and to keep in their good books."

Narratives of Wetland Science

Complementing these polar bear stories, a large part of my fieldwork in 2002 was spent as a field assistant for a project at wetland sites near Resolute. This involved "commuting" by ATV from PCSP Resolute to field sites in wetlands across Cornwallis Island.

My initial entrance to this site occurred because a student, Claire, needed to stay at the PCSP base one afternoon in order to do some packing for a field trip to Cresswell Bay before the female lead professor returned from southern Canada. This meant that her ATV was free, and Dominic, the other student researcher, asked me if I would mind helping him out.

4.3 | Hydrologist in wetland, near Resolute, July 2002

The wetland site, S1, was about two kilometres away from PCSP Resolute. The project involved measuring hydraulic conductivity at the site, which required taking measurements at a series of points (figure 4.2). When moving between the measurement points, we had to walk in straight lines "to minimize damage." We also took soil samples at S1, as well as at another site, S2, around 200 metres away.

It was apparent that this sort of field science is very monotonous, involving the daily performance of repetitive tasks by the student field assistants. I helped Dominic on about twenty occasions. Often we went out to the sites after dinner, working until after 10 p.m. to take advantage of the optimal daylight.

As it was near the beginning of July, the permafrost was melting, and this made it increasingly difficult for Dominic to get around to his measurement points (figure 4.3). It also meant that ATVs got stuck in the mud. Even when it was not snowing, it was frequently very cold, foggy, and muddy in the field. The problematic environmental

4.4 | Hydrologist using ATV to move between field sites, near Resolute, July 2002

conditions also appeared to frustrate Dominic. On a few occasions, and prior to recording water quality, Dominic drove his ATV up a stream flow in order to reach a soil sample point (figure 4.4).

Some important issues were highlighted during this fieldwork. On a particular afternoon, Dominic lost the cap for the pH meter. This was important, because without the cap, the pH meter "would dry up." We looked for it for over an hour in the wetland. This was a difficult situation, because it was not the research group's pH meter. It had been borrowed from another group because Dominic's original pH meter was "not calibrated." Dominic therefore faced a stark choice. Explain to the other group that he had lost the borrowed cap, or give them his group's cap, and then be reprimanded by the lead professor when she arrived from the south. This showed the importance of co-operation between *some* research groups, as well as the issues that relatively powerless student researchers felt they had to deal with in the field. At the same time, it indicates the critical importance of junior members of field teams to the overall success of a research program.

During one afternoon's research practice, Dominic mentioned that he was interested in how the wetlands are affected by the footprints of researchers. Such footprints form puddles around pipes, wells, and piezometers. For this reason, Dominic stated, "I would love to do an MSc on the human impacts upon research."

Moreover, being away from the base provided an opportunity to talk through many local issues. The field sites often acted as a space of dissent for scientists to air views they could not discuss around the base complex. At lunch one day, for example, many students of a particular professor complained about receiving inadequate instruction on exact practices to follow in the field, and of a failure to be kept informed about how their fieldwork fitted into a larger research project. Later that afternoon at the wetland, Dominic reflected that he believed that this situation was because that professor, not his, lacked sufficient research funds to hire two field assistants and to purchase and maintain the proper equipment. This was why, Dominic believed, the students of the other professor felt over-stretched and were so disgruntled.

Fly-Camp Science in Action

Given their respective locations across the High Arctic, the fly-camps were much harder to access, but I was able to observe field practices at a number of sites. This was crucial to grasping the composition of practices in Arctic field science. As Quentin Lefebvre succinctly put it, "In the office, field camps are dots on the map. It's different when you're in the field. It's at the camps that you'll get a real feel for the frustrations, when the radio is the only contact with Polar Shelf, and you know that the Twin Otters are not coming for you, but you can hear their broadcasts as they fly somewhere else."[26]

In 2001, I helped evacuate a camp at Cresswell Bay and did fieldwork at Mould Bay. In the following 2002 season, I helped establish a fly-camp at Stenkul Fiord, on southern Ellesmere Island, and helped evacuate another on Truro Island. However, the best access to scientific activities at a fly-camp was provided by a period of residence at the Quttinirpaaq Parks Canada base at Tanquary Fiord, northern Ellesmere Island (figure 4.5).

4.5 | Parks Canada base, Tanquary Fiord, Quttinirpaaq (Ellesmere Island National Park), July 2002

I met Rex Driver, from Parks Canada, on the penultimate day of my 2001 season, as his group returned from Ellesmere Island at the end of their summer residence. Rex believed that PCSP had emerged from the Defence Research Board. When I suggested that there might be a more complicated history, he invited me up to the park during the following field season. Rex was accompanied by a summer student, Terry, who believed that PCSP was "a boring topic." As he asked, rather demoralizingly, "Why are you studying that? Doesn't it just provide logistical support?"

My initial purpose was therefore to give a presentation on the history of activities by PCSP and related organizations on northern Ellesmere Island. However, as there are still PCSP buildings and PCSP equipment at Tanquary Fiord, this was also an opportunity to get a better grasp of the extent of the former operations of Polar Shelf.

On 15 July 2002, I flew up to the Parks Canada base at Tanquary Fiord, together with Pierre, a French-Canadian parks employee, Kirk, a Parks patrolman who was a summer employee from Resolute and was heading on to Lake Hazen, and Julian, from Water Survey Canada, who was doing some "contract science" for Parks.

For my time at Tanquary, the base staff comprised Rex, Pierre, Terry, myself, and Joseph. Joseph, a sixteen-year-old summer student from Resolute, was the younger brother of Kirk, whom I had met on the flight up. I used the time at Tanquary to examine the Polar Shelf buildings and observe how the Parks staff fulfilled their commitments to conducting various types of monitoring science. I also helped with duties around the base. All the staff, except Joseph, took turns to cook meals.

Evening activity around the base included walks in the astoundingly beautiful surrounding area, but for the long-term residents it focused on DVDs that they borrowed from the military at Department of National Defence (DND) Alert. Military helicopters frequently flew past Tanquary Fiord as part of Operation Hurricane, a refuelling operation for the military micro-meteorological stations between Alert and Eureka. These helicopters often stopped at Tanquary and sold cans of soft drinks, or loaned DVDs, to the Parks staff. Incidentally, the Parks staff were not allowed alcohol on base, and they adhered to this rule rigorously.

After my presentation on the history of scientific activities on northern Ellesmere Island, the Parks staff were all very grateful, but saw little importance for PCSP beyond the provision of logistics. For Rex and Pierre, this support is still critical to the National Park, as it allows cheap and efficient provision of flights, accommodation at Resolute, and radio communications through the "economies of scale" available to PCSP.

The scientific practices undertaken by Parks appeared to be slightly chaotic. Pierre had flown up in order to download data from automated weather stations, but he had a problem with most of the loggers and feared that they would lose an entire year's worth of data. This had apparently also happened in the previous year and thus there was the potential loss of the data from twenty-four months.

Moreover, the Parks staff were preparing to undertake fieldwork on Ward Hunt Island at the end of my week of residence. Rex and Terry were doing much packing and repacking, in order to get the load weight down. On the flight on which I was to return to Resolute, further Parks staff were first coming up to Tanquary for the Ward Hunt fieldwork. The group thus had to pack for six people for ten days, and the maximum Twin Otter load is 2,200 pounds. One person with individual gear is estimated at 200 pounds. This left only 1,000 pounds for all other food, supplies, and scientific equipment. However, when the new Parks staff arrived, the Twin Otter was already overloaded, as the group coming up had not reckoned on these calculations.

The Parks staff therefore appeared relatively disorganized. Examples of this confusion were legion. Pierre was desperately trying to set up his weather station on the flight up from Resolute to Ellesmere because he thought that there had been a change of plan and that he would be going straight out into the field at Fort Conger. This schedule changed again when we arrived at Tanquary. On disembarking from the Twin Otter, Rex had forgotten that I was due on that day and had lost a spirit-level that was necessary for Julian's field research.[27] Rex had also forgotten to book a helicopter for Julian, which meant that he had to cram a week's monitoring research into one afternoon. The plans thus kept changing but, as Rex put it, "That always happens up here."

Unlike most federal operations in the High Arctic, Parks appeared to have a relatively high proportion of Inuit on staff. As well as Joseph and Kirk from Resolute, Rex had an Inuuk mother, and Terry was fluent in Inuktitut, having been brought up in Igloolik. There were many discussions on base about the politics of Nunavut and Inuktitut language education. At the same time, there was a little social tension, as one employee from Ontario, with whom I overlapped on base for only a few hours, was very dismissive and patronising to Joseph. There were constant remarks, even in this very short period, about him being lazy, and I detected a hint of racism. Later, Rex informed me that he had "had to tell her to leave Joseph alone as she's not his boss."

Being a Field Person

From these observations of practices around PCSP Resolute and at other field sites, it is evident that there is the general construction of the notion of being *both* an accomplished scientist *and* a good "field person." James, for example, was an increasingly grumpy biological scientist towards the end of his time in the field. At one mealtime, he remonstrated, to a captive audience, about the importance of the Sony Walkman during fieldwork: "They help you forget someone is there.

If you are in a tent with just one person, who is not your wife, then it grates, inevitably. A Walkman helps you disappear." This was evidence to me that James was not a good field person, as Roots and an older generation of PCSP scientists would have it.

Aptitude for Arctic fieldwork, despite the changes in technologies, everyday practices, and gendered subjectivity, is still demonstrated, in the main, by being a "good ice man." Leif Lundgaard, a former base employee, is still a Polar Shelf legend. As Walder put it, he was "a *real ice man*, who used to love sleeping outside." He once "fell asleep on the ice, after drinking a few, and woke up with his beard frozen to the side of the Parcoll tent." Walder had to force Leif to retire in the late 1990s, "because of his age, though he is still fitter than most twenty-five-year-olds." Andrew said I must meet Leif, because "he's the field talking, not the politics."[28]

It is not only through these tales of retired PCSP staff and scientists that the importance of the good field person is perpetuated. During one conversation, Andrew Walder asked me, "Where were you in March and April? Those are the *real* scientists, out in -40°C. Not these fly-by-night summer students." These supposed standards regarding the capability to cope with the environment were often deployed in power struggles between the staff on base. Thus PCSP staff might regale scientists about other staff being unable to cope with some situation in the past. These might include failure to deal with the weather or isolation, or using the Arctic to escape from problems in the south.

Many scientists articulated a similar view of themselves as Arctic fieldworkers, though usually in jest. Nicola, whilst knitting on our flight back from Mould Bay, commented, "Now you see what it is like to do Arctic fieldwork. We are a rare breed." Similarly, Quentin Lefebvre discussed the cold and the inconvenience of Arctic fieldwork but noted, "You can't get such an experience doing science anywhere else. The exotic nature of the Arctic is a significant drawing card for all of us northern scientists."

Numerous interviewees argued that PCSP had a crucial role in establishing the safety procedures for northern science, and obviously such considerations are important. The base managers argue that if a camp misses two radio skeds, then PCSP will still technically send a flight out to get them. In reality, this is always a judgment call, and small camps of two people or an inexperienced camp would be treated with greater caution than large camps of six people under a senior field researcher.

However, the most obvious way for scientists to demonstrate that they are a good field person during the summer, when the weather is much less harsh, is to *appear* to be blasé about the polar bear threat.[29] When Alan Merton was initiating a project on Melville Island, where he had not worked before, Andrew was stressing to him the high levels of polar bear activity in the area. Merton was very surprised and had obviously not prepared for it, but he was not particularly anxious. However, eventually Merton was persuaded to take an extra ATV because of the potential threat. This meant, because of the payload, leaving an igloo tent behind and staying instead in the huts built by the Canadian Wildlife Service and Government of the Northwest Territories. The party also took two .308 shotguns and borrowed extra bear spray from PCSP. The extra ATV also meant that the overall payload had to be decreased, so the group spent the night before going out to the field repacking their food and supplies. Logistical decisions therefore dictate everything, even questions of safety and subsistence, for a good field person.

Moreover, lest anyone assume that these constructions of field identity can be reduced solely to male fieldworkers, as a sort of naive machismo, I also observed them being articulated by women towards other women. A female professor was angry at a student's behaviour during her absence from Resolute. The professor had heard that the student was drinking too much, and believed that "she can't drink every day because she'll ruin her master's." Moreover, for other physical reasons, "she should not be doing fieldwork. She is excited by the drinking at Resolute, *not* by the science. If she does a PhD, it should involve her sitting in front of a computer. It is expensive to stay up here, and [name withheld] told me that [name withheld] spent all her time drinking last year. I am worried, because I have spent my time since 1983 building up my reputation, and having this student reflects badly on it." For a whole host of reasons, then, it appeared that this student was not a good field person, and this hampered her ability to conduct a successful scientific project. Articulations of being a good field person, then, circulate across the cultures of PCSP.

Conclusion

Much can be learned from these observations of field practices at PCSP Resolute. The actual quotidian activities of current Canadian scientific projects in the High Arctic have hitherto been ignored. Through detailed records of the "complex specificness" of my fieldwork (Geertz 1973, 23), I have tried to show that the PCSP, through its base managers and establishment of rules of support, continues to perpetuate a vision of the relationship between Canadian nationalism and scientific practice that has evolved from the founding of the PCSP under the Diefenbaker governments of 1957–63. These rules help constitute practices at the base complex and in the respective field sites, by making scientists determined to achieve the precarious status of a good field person in a culture of temperance. It is impossible to understand why scientists use a particular field site, or mode of transport, or instrument, without grasping the rules that structure PCSP support. Such understanding can emerge only through *ethnographic* research, which allows the description of practices within an interpretive context.

At the same time, through the observation of individual actions through my fieldwork, I have begun to suggest the importance of viewing human agents as emotionally competent and reflexive beings. In chapters 6 and 7, therefore, I begin to amplify these emotions and voices across different levels throughout the institutional spaces that constitute the PCSP. In doing so, I develop my argument for the analytical poverty of research attention focused entirely on practice and for the continuing need for *interpretation* by the researcher in order to gain understanding in the field.

5 Canada Day in Qausuittuq: Dramatizing Inuit Encounters

The insistence by Inuit and other minorities on cultural difference that is meant to empower can, however, marginalize and isolate.

Pamela Stern, 2006a, *Critical Inuit Studies*, 263

Introduction

The Arctic regions have often been conceptualized in the Western imagination as antagonistic spaces where pathological natures are to be overcome. The unrelenting physicality of both passage and dwelling, combined with spatial isolation, contribute to the ideological resources required to construct meanings for such landscapes. Indeed, it is the valorization of the ability to meet this corporeal challenge that has historically encouraged visitation from non-indigenous peoples. Moreover, polar sublimity continues to circulate through global discourses. Recent debates over climatic changes and their impacts upon charismatic mega fauna such as polar bears, for example, have focused international attention on the Circumpolar Arctic.

But as many commentators have noted, for all its universal attraction, the Arctic is also *home* to the Inuit. The Arctic is *Nuna vut* – Our Land. In Canada, the political instantiation of this is evident in the settlement of the land claim between the federal government and the Inuit, resulting in the creation of the Territory of Nunavut on 1 April

1999 (Bravo 2000). With this recrafting of political cartography, the implicitly colonial depiction of the Arctic as an uninhabited wasteland has been recast as an episode consigned to histories of empire. As such, the region has begun to provide an exemplar case for studies of de-colonial resistance and indigenous self-determination. Through force of will and organization, the long reparation and rehabilitation between Western states and northern indigenous peoples has, at least, commenced.

As might be expected, the territorial instantiation of Nunavut notwithstanding, the relationship between Inuit and Canada remains volatile. The multiplicity of colonial presents in the contemporary world have variegated geographies (Gregory 2004). In the Canadian Arctic, histories of colonial dispossession and state intervention have been resisted through indigenous agency. This chapter shows how such cartographic refashioning can have remarkable consequences for those peoples beginning to achieve control over their own lives. As newly enfranchised members of the precarious Canadian nation, Inuit have begun not only to discard, but also to reformulate and even to embrace, elements of the wider national culture.

What is most important to remember here is that Arctic lands are *spaces of encounter*. The Arctic has witnessed histories of interaction between different groups: explorers and missionaries with Inuit; scientists with indigenous peoples; even, more recently, between scientists and adventure tourists. What holds these disparate encounters between Inuit and Qallunaat together around a common theme is that all have revolved around power relations.[1] Feminist anthropologist Sherry Ortner has shown, in work on the history of interactions between sherpas and Western mountaineers in the Himalayas, how encounter is fundamental to social histories of *both* communities (1999, 2006). As Ortner argues, in contexts of encounter "what is at issue are the ways in which power and meaning are deployed and negotiated, expressed and transformed, as people confront one another within the frameworks of differing agendas" (1999, 17).

This chapter uses the notion of encounter as a point of departure to think more seriously about the Arctic-as-homeland. Whilst domesticity and colonialism are persistently enacted, the Arctic can also be thought of as a *social* space. That the power relationships underpinning these interactions have historically involved racism and epistemic violence, and will likely continue to do so, should not distract us from examining cross-cultural encounters in the ethnographic present. Human actors interact in the Arctic and, as they do so, they perform important social practices. In other words, Arctic spaces can be conceptualized as involving social formations that must be analyzed not as inherently pathological but as, in many ways, commonplace.

Despite this apparent need for understanding the social Arctic, it was not until the 1970s that anthropological attempts were made to document contemporary spaces of encounter in Canada's North. Research by Hugh Brody and Robert Paine uncovered the sociology of settler communities in northern Canada. As well as power relations positioning Inuit in subordinate roles, Paine (1977) uncovered the malign impacts upon both communities of the consistent expectation of tutelage relationships between whites and indigenous peoples. In an elegant ethnography of Iqaluit, Brody (1975) was able to demonstrate more starkly the destructive tensions evident between settlers and Inuit during the early 1970s.[2]

In more recent decades, the Canadian Arctic has become an arena of contestation in debates over global climatic changes. Encounters between settlers and Inuit have thus begun to be supplemented by those between Inuit and scientists. Such interactions result in social positionings that continue to produce apprehension within both communities.

In what follows I use notions of ritual and carnival to explore how Inuit relieve community tensions and negotiate identity in spaces of encounter. Drawing on theories of ritual developed by anthropologist Victor Turner, the chapter pursues the enactment and alleviation of social tension. This is accomplished by drawing from a single ethnographic interlude – the performance of Canada Day in Resolute in 2002. For Resolute, the site of a federal scientific field station operated by the Polar Continental Shelf Project, as well as the second most northerly community in Canada, possesses a controversial history of encounter between Inuit and Qallunaat.

Lest I be misunderstood, it is worth restating that this book is not aiming at an ethnographic depiction of the Inuit community in Resolute. Rather, this chapter deploys an ethnographic account of encounters between scientists and Inuit during one particularly significant moment during field research. However, before discussing these events, it is necessary to expand upon Turner's understanding of ritual.

Theories of Ritual, Carnival, and Social Dramas

Anthropologist of performance Victor Turner construed societies as constantly enacting social dramas. Influenced by the Manchester School of Social Anthropology associated with Max Gluckman, Turner saw such dramas as performing "'classificatory' *oppositions*" among communities (1982, 11; original emphasis). Social dramas allow communities to identify conflicting groups and for individual members to distribute associations accordingly. These dramas are both episodic and pervasive across human activities. As Turner puts it, "Social life, then, even its apparently quietest moments, is characteristically 'pregnant' with social dramas" (11). For Turner, this theorization was exemplified through the example of the Rio carnival. The carnival, as the apogee of consumptive excess and through the inversion of standard social roles, enables the *relieving of the pressure of existing social conflicts* (Turner 1987, 1974).

The theory of the carnival, of course, owes much to Mikhail Bakhtin's study of Rabelais (1984). For Bakhtin, carnivalesque moments enact the inversion of quotidian social hierarchies. In short, they form instances of the everyday world being "turned upside down." As Stallybrass and White argue, the conversion of Bakhtin's notions of carnival and parallel trends in symbolic anthropology was particularly significant (1986). Across both the social sciences and literary studies, scholars became interested in the subversive potential within instances of quotidian life. Anthropologists such as Sherry Ortner interrogated societal rituals in order to reveal wider cultural processes (1978). More recent work by geographers has included investigations of gender performance at the Rio carnival and the politics of the annual Caribana festival for the Afro-Caribbean community in Toronto (Lewis and Pile 1996; Jackson 1992).

However, here I am less interested than Turner in attempting to decipher social structure from such dramas. Rather, I want to highlight the notion that communities enact rituals as a way to relieve episodic social tensions. Participation in such carnivalesque moments, I argue, allows the inversion of power relations within both everyday life and the broader Canadian state.

In this chapter, I demonstrate that the performance of the Canada Day celebrations in Resolute achieves such a sociological function.

This is accomplished by discussing the Canada Day ritual as a drama of social actors. To situate such a discussion, however, it is necessary to outline the historical geography of Resolute. In doing so, it should become apparent why the performance of Canada Day at this site is of such ethnographic note.

A History of Resolute: Communications Hub of the Canadian Arctic

The hamlet of Resolute, Nunavut, is located on the southern edge of Cornwallis Island (Powell 2005). Archaeological traces of Thule encampments dating from over 300 years ago have been found on Cornwallis and neighbouring islands (McGhee 2001). However, the location of the Inuit hamlet and scientific research station at Resolute was fortuitous. Indeed, there was never any formal intention to develop a modern presence at Resolute at all (Tester 2005).

Although Cornwallis Island was first visited by Edward Parry during his search for the Northwest Passage in 1819, the hamlet of Resolute Bay was named after HMS *Resolute*, a ship abandoned by the British expedition under H.T. Austin during 1854 (Tester 2005). It was not until the Second World War, however, that extensive twentieth-century human activity was evident in these High Arctic regions of Canada.

Geopolitical considerations after 1945 had led to discussions between the Canadian and US militaries over the need to establish meteorological bases and maintain a presence in the High Arctic. In February 1947, an agreement was reached to establish five Joint Arctic Weather Stations (JAWS) in the region. These bases were to be constructed at Mould Bay, Eureka, Alert, and Isachsen with a central station at Winter Harbour on Melville Island. However, in summer 1947, during the sea mission to construct the supply station at Winter Harbour, the propellers of the icebreaker USS *Edisto* were damaged in heavy sea ice. This resulted in the remedial choice of an alternative site on nearby Cornwallis Island.

Given this hasty adoption of Resolute Bay, a number of unanticipated issues were to emerge with the harbour as a landing site for sea traffic. The site was ice-free for a season of only seven weeks and possessed shallow landing beaches. There were also further complications resulting from drifting sea ice. Such problems notwithstanding, given the difficulties in establishing any such station from the outset, the Royal Canadian Air Force constructed an airstrip at Resolute in 1949, thereby consolidating its status as the communications hub of the High Arctic. In the following decades, the airport acted as a central co-ordination site for scientific and military activities by the Canadian government. During the massive expansion of natural resource exploration over the 1960s, Resolute served as a logistical centre and saw the growth of commercial operations. The Canadian military maintained control of Resolute airport until 1964, when management was transferred to Transport Canada. When the United States withdrew support for the JAWS in 1970, the meteorological station at Resolute was maintained as a High Arctic Weather Station by Environment Canada.

As this discussion suggests, the connection between Canadian political sovereignty and the site of Resolute has remained since its earliest emplacement. These military and scientific presences were to be exacerbated by resettlements of Inuit to Resolute.

A History of Resolute: A Relocated Inuit Community

In the summer of 1953, three Inuit families from Inukjuak (Port Harrison), Quebec, together with a family from Pond Inlet, were resettled at Resolute Bay, following the recommendation of the Federal Department of Northern Affairs and National Resources. At the time, the stated reasoning for relocation was the alleviation of pressure from hunting on the ecosystems of northern Quebec. By the early 1950s, the area inland from the east coast of Hudson Bay had supposedly suffered a serious decline in game, particularly affecting caribou herds (Robertson 2000). It was thought that the Inuit would be able to provide a seasonal labour force for the airport and central bases at Resolute. A small Royal Canadian Mounted Police (RCMP) detachment was also deployed in Resolute to administer the relocated families.

Twenty-two Inuit were transported to a region beyond their imagined and practical geographical landscapes. As perhaps might be expected, major socio-economic hardship was encountered by the Inuit families at Resolute Bay. Meteorological records since the 1940s indicate that Resolute consistently has colder temperatures, more precipitation (as snow), and lower visibility than almost anywhere else in the Canadian Arctic. The environment and winters were considerably harsher than those encountered in northern Quebec, resulting in problems of inappropriate clothing and nutrition. This was further complicated by the complete diurnal darkness encountered at Resolute for over five months every year. Inuit came to know their new home as *Qausuittuq*, or the "Place where dawn never comes." Moreover, by being relocated outside normal hunting areas, the families were often unable to use traditional skills and practices. Many Inuit were simply left reliant on scavenging from refuse left by the base complex.[3]

A number of social problems also resulted from the relocations. Inuit women were effectively confined to the community, because the RCMP disapproved of possible relationships developing with the military and civilian personnel stationed at the base complex. The RCMP effectively facilitated a social-spatial separation of the hamlet from the base. In instances where cross-cultural encounter did occur, these often revolved around the airport bar. This resulted in problems of alcohol and suicide within the community.[4]

Through the 1960s, further Inuit migration into the community resulted in the town site expanding directly under the airstrip approach. The community was therefore transferred to its present location in 1975. This site is eight kilometres from the base complex at the airport. Connected only by a single-track road, this has reinforced the separation of the two populations.

The major reason for the relocations appears to have been part of an attempt to demonstrate Canadian sovereignty over the High Arctic. It was believed that Inuit, being indigenous to northern Canada, helped sustain political claims by the Government of Canada to the region, given perceived threats consequent to increased military presence by both the United States and USSR during the early Cold War. This was obviously problematic from its very inception, given that the Quebec Inuit did not possess traditions adapted to the environments of the High Arctic. However, the institutional mindset of the 1950s was predisposed to this sort of welfare colonialism, believing that Inuit were being relieved from suffering caused by lack of success on hunting grounds around Inukjuak. It should also be noted that members of the administrating Department of Northern Affairs and National Resources have continued to state that the Inuit families were fully supportive of the relocations.[5]

In the late 1970s, some Inuit returned to Inukjuak at their own expense. This led the Makivik Corporation, a group established by the Inuit of northern Quebec (or Nunavik), to seek compensation and an

5.1 | Hamlet of Resolute Bay, August 2001

apology from the federal government in 1982 (Tester 2005). Following the establishment of the Royal Commission on Aboriginal Peoples (in 1991), the survivors of the original resettlement from Quebec were awarded $10 million compensation in 1996 after the publication of the report.

Resolute hosts a population of approximately 215, of which 79 per cent were reported as of "aboriginal identity" in the 2001 Census of Canada. This population is composed mainly of Inuit, a detachment of RCMP officers, and a few other federal and territorial employees (such as teachers), and periodic visitors (figure 5.1).

This altogether too brief diversion into the history of Resolute is important, as it indicates that the relationship between the Canadian state and the Inuit of the hamlet is fraught and contested. The question remains as to what the consequences might be for Inuit identity in contemporary Resolute.

Reformulating Inuit Identities and the Creation of Nunavut

As discussed in chapter 1, there have been attempts to create a pan-continental national identity in Canada since, at least, Confederation in 1867. The tense history between the two founding nations of French and English Canada has heavily influenced debates around Canadian national identity, but these have been reinvigorated in recent years by notions of multiculturalism (Taylor 1991; Kymlicka 2001, 2003; Harris 2001). In short, issues of bilingualism and asymmetric federalism have come into contest with those of indigenous self-determination and immigration.

What is important for our purposes is that in much of this discourse around nationhood, Inuit have come to occupy the apex of a refashioned Canadian identity (Hulan 2002). For example, for political commentator John Ralston Saul, First Nations peoples form a critical strut of a tripartite Canada, together with anglophones and francophones (1997). But Inuit, because of the importance of northern environments in this imagined geography, form the core of the "positive nationalism" required for Saul's Canada (508). As Saul concludes his extensive essay on nationhood and Canada's place in a globalizing world, "The Inuit quality of *isuma* summarizes that essential context. It has as much to do with positive nationalism as with the public good. *Isuma* – intelligence that consists of the knowledge of our responsibilities towards society. It is a characteristic which grows with time. If you choose to look, you can find it at the core of events through the long line of Canada experience. It is an intelligence, the Inuit say, which grows because it is nurtured" (508).

Inuit traditional knowledge and practices have thus come to be seen as fundamental characteristics of the Canadian nation. This remarkable reversal of fortune needs to be seen in the context of the creation of the new Canadian territory of Nunavut in 1999. As Inuit activist John Amagoalik remarks, it was not so long ago that journalists travelling north "almost never failed to refer to the Arctic as a 'wasteland where nobody lives'" (2000, 138–9).

Following the resolution of the land claim between the Inuit and the Government of Canada, the Nunavut Land Claims Agreement was signed in May 1993. This agreement provided for the creation of Nunavut and the establishment of a public government for the governance of this new territory. Although Inuit would form the majority of the population, around 85 per cent, it was not an indigenous political structure (Hicks and White 2000).

The Government of Nunavut therefore allows for a degree of indigenous self-determination within the structures of the Canadian state. As might be expected, the formation of the Territory of Nunavut has had complicated impacts on Inuit conceptualizations of identity. For some, the rehabilitative political processes involved in the creation of Nunavut will not be complete until Inuit Qaujimajatuqangit – or traditional knowledges, skills, and ways of thinking – have been completely incorporated into the government structures. *Nunavut* might mean "our land," but the bureaucratic structures of the fledging territorial state are still very much those of Qallunaat.

Like all identities, the Inuit sense of self is complicated. One of my Inuk informants from Pangnirtung had a Qallunaat father and an Inuuk mother. During conversations, he revealed that he had suffered racism, as he put it, "from both sides" during his childhood, but especially from other Inuit. His mother had kept him away from her Inuit family because of social problems endured by the community through the 1970s and 1980s. Yet he feels firmly that he is Inuk. Furthermore, he claimed that it is sad that "in Resolute and Grise, Inuit commonly speak English, not Inuktitut." This is ostensibly different from the situation in other settlements, such as Igloolik and Pangnirtung, where Inuktitut is commonly spoken across the community. In short, there is a geography of the practices of Inuit identity within the new territory of Nunavut; and in Resolute, being Inuk is particularly complicated.

Traces of the State: Federal Presences in Resolute

Across Nunavut, apprehension remains for Inuit in the face of any manifestation of the federal state. These stresses are exacerbated by both the permanent and the seasonal presences in Resolute. The current federal contingent in Resolute includes a two-person RCMP detachment, a minimal Environment Canada staff at the weather station, a small Transport Canada staff at the airport, and the large seasonal population of the Polar Continental Shelf Project.

One evident tension is the manifest hostility expressed by Inuit towards the RCMP. An adolescent Inuk informant from the hamlet told me that he did "not like the police, because of what they did in the old days." These stories had been passed on to him, like traditional hunting and navigation skills, as part of his becoming Inuk in Resolute. However, the small RCMP detachment is located within the community. It is the undertaking of scientific research through the support of the Polar Continental Shelf Project, based out near the airport, that provides the major influx of people into Resolute during the summer months.

This presence has an important impact on the long-term sustainability of the Inuit community, not least by making viable arrivals/departures by jet from Ottawa and Edmonton every week during the summer. However, as Ralph Alexander of the Resolute Hamlet Council puts it, there has been an evolving relationship between the scientific base at the airport and the local community.

> It has changed an awful lot. Back in the seventies, you had the base and you had the village. And there was a bar, so everybody, sort of, knew everybody else. And the bar closed in '81, which may have been a good thing in one way – a little less drinking. But then in another way, it meant that people didn't meet each other anymore. And by this point in time there are people at the base that nobody here even knows about. But at the same time, twenty years ago the base did what they were going to do and you did what you were going to do. Now … there's more regulation of what happens. So you have a better idea, at least, of some of the management area, the higher levels. So you know more or less what's being done by Polar Shelf, what field parties are going out where. To ensure that they're not going to cause any problems for people who actually still hunt.[6]

As discussed in chapters 3 and 4, weather conditions have tended to result in a flexible approach to research plans amongst the Arctic scientific community. If pilots are unable to land at one site, traditionally a nearby site is chosen, and research plans are adapted. This is an important part of Arctic scientific practice that was frequently stressed by researchers during interviews. However, stricter licensing procedures in Nunavut have made this adaptability increasingly difficult to maintain.[7]

It would appear, at least, that attempts to manage research through new licensing arrangements operated by the Nunavut Research Institute in Iqaluit have allowed Inuit, or at least their political representatives in local government, more knowledge of activity on their territories. This means that the ostensible view of the PCSP within Resolute is relatively indifferent. There is suspicion of the activities of scientists, but potential interference with wildlife or hunting grounds is in decreasingly serious dispute. As Alexander argues, the current view of PCSP by the inhabitants of Resolute is generally aloof:

> It's kind of neutral. I mean there's always going to be somebody saying, "Well, they're going up, and they're just surveying the animals, and that's not good for us." And there's like the other thing, "Well, they're coming up, and what they're doing isn't bothering anything, but like, we don't get anything out of it." And, you're kind of saying, "Well, look, there are people being employed out of it, so there is some benefit." I think you're going to see that anywhere you go. I think the biggest thing is, when you don't know, you think that maybe what they're doing is hurting your hunting or causing problems with the animals. And once you see what they're doing, it's usually of very little interest … in general. Because they're not hurting the wildlife. You've got two people sitting in a tent on Ellesmere Island for a week taking water samples … The only ones that really concern anybody here is if they are doing helicopter surveys. Is it going to go into wildlife areas?[8]

This indifference notwithstanding, there are still major interdependencies between the base and hamlet communities. A major way in which the base population and the hamlet interact is through employment. Indeed, the potential provision of seasonal labour by Inuit was an initial justification for the indigenous relocations to the region. However, as discussed in previous chapters, many base jobs are performed by migrant labour from Newfoundland. As Alexander argues, though, in many ways the lack of reliance on seasonal labour opportunities with PCSP is actually an indication of the relatively high levels of employment in Resolute. "I think it has always been a fairly high level of employment here. In a lot of the communities you had very high levels of unemployment. Here, you don't. Now, a lot of the jobs here are Co-op or the Hamlet. And there's two hotels in the community, so there's employment there. And there's a third hotel up at the airport. Polar Shelf has never employed a lot of people from the village. But they do employ one or two every year. In fact, they employ them through the Hamlet. It saves them paperwork."[9]

Nevertheless, although most of the PCSP staff are from southern Canada, this Inuit labour is crucial to the function of the organization. Rachel and Emma, part-time cleaners at the PCSP base, and Alistair, who drives the water truck and does some manual labour on an "odd-job" basis for PCSP, are all from the hamlet of Resolute. One of the base cooks, Sarah, is from Inuvik in the western Arctic. Every Sunday afternoon, Emma and other Inuit women from the village, and sometimes their children, visit PCSP to talk to Rachel and Sarah in the dining room at the base. Topics for discussion include issues such as regional and dialect differences among Inuit.

Most scientists and base staff were respectful towards Inuit and their traditions. For example, Edward, a base manager, was largely sympathetic to attempts, as he put it, to get "more natives involved in science," and to the Nunavut Research Institute's apparent desires to develop

> something like Polar Shelf, but that is not just another federal presence. You could develop the expertise in the communities for this Nunavut decentralization plan [for Arctic logistics] to work. But it takes time to develop that expertise in every community. Borek and First Air have three or four guys in lots of communities. They can't just have one, in case someone goes out on the land. But the communities are much more easily persuaded to see the importance of a plane company, because they bring in food and relatives when they visit. It is much more difficult to persuade them of the importance of science. It is possible, but we need someone on the ground to go round the communities persuading them. And Nunavut needs to put the money in if it wants to do that.[10]

Another major interaction between the base and the community is through the employment by field parties of Inuit guides as security against threats from polar bears. Strict regulations on the size and composition of field parties supported by PCSP mean that research teams will often hire a hunter to join the party for a period in the field. The stark seasonality of this Inuit labour with PCSP means that it is combined with other wage employment and subsistence activities. As Alexander states, during July 2002, various individuals helped in scientific activities in the field without any direct recording of their presence.

> They [PCSP] have three people from the village working. There's somebody who works with them every year, and then there's a couple of people that they hire on a … on a need basis … But the other people that they have working for them right now are working in town … One person is paid through the Hamlet, one person is a direct employee. However, we're not really an employment agency. Somebody may call me up. In fact, they do call up and say, "Well, we're going out somewhere. Can you give me some … a couple of names of people that might be interested?" And we'll give them two names, and we'll give them phone numbers. And it's up to them. And we don't even know if they go out or not.[11]

Other quotidian interaction between scientists and base staff at PCSP and the local community remains limited. Such encounters involve occasional economic transfers at the local Co-op store. Scientists and base staff will make trips to buy soft drinks and sundries from the store whilst on base.

Despite the general low profile of cross-cultural engagements in Resolute, some colonial views of Inuit still persist. Such traces of the colonial present are often covert and become manifest in banal ways. This is evident, for example, in the pervasive use of the term "locals" by the PCSP base staff and the majority of older scientists. Research groups, especially those working out of Resolute, spoke of hiring "a local and a dog" to provide assistance in the field. However, younger scientists, and especially students, appeared to offer greater recognition by using "Inuit."

But at PCSP Resolute, it appeared that this lack of serious understanding occasionally resulted in tension between the communities. Nunavut Day, celebrated annually on 9 July, is a public holiday celebrating the passing of the Nunavut Acts. Consequently, it is a paid holiday for territorial government employees, and most stores are closed. Outside of Nunavut, perhaps understandably, Canadians tend not to mark this day, and this can result in difficulties. On the morning after

Nunavut Day, a pilot entered the base office to complain that the Post Office (Co-op) in Resolute was closed. Andrew Walder's response was to explain that this was because the Co-op was having problems keeping its staff, as everyone kept quitting. As he joked, "That's life in the North. You go to a restaurant for lunch, and there's a sign saying 'Gone for lunch.'" It may simply have been, of course, that the staff were enjoying the repercussions of what had been a public holiday for them. Later that morning, Andrew rang Rachel to ask why she had not shown up for cleaning work at the base. Rachel tried to negotiate a later start time at 1 p.m., but Andrew refused and insisted that she come in at 11 a.m. I knew from previous conversations that Rachel had been attempting to get "this day off," I assumed to recover from the partying of the night before. However, most southern Canadians on base were completely oblivious to the importance of the day to Inuit.

Amongst some Canadian scientists, there was a general lack of empathy with Inuit society. On one afternoon, a group of scientists made a trip to the village in the hope of buying some artifacts from a distinguished local carver. The carver was obviously asleep when they arrived, but the scientists continually knocked on his door, assuming that he should be awake at this time of day. During the constant summer daylight in Resolute, members of the community appear to sleep little, and when they do they tend to nap for short periods in the morning or mid-afternoon.[12] A young undergraduate scientist, after buying an artifact, forced the reluctant Inuk to pose for a photograph with her, and then commented that Inuit left everything strewn around their houses and that this made the village appear ugly and untidy. She wondered aloud, "How do they live like this?" Another older, female scientist in the group wondered why the Inuit mothers allowed their children to just play out in the streets of Resolute, when there had been warnings of polar bears in the area.

However, the worst example of a lack of cultural sensitivity was provided by a Kenn Borek helicopter mechanic on contract with the PCSP. The mechanic wanted to fly home, but he had to stay to maintain the helicopter until the end of the season. However, the managers would not let him stay at the base anymore, as he was surplus to PCSP requirements. Consequently, the mechanic had moved to the South Camp Inn in the village. This appeared to be the source of his discontent, as he felt the inn was "not like a hotel. It is like staying in someone's house who you don't know. It's hell and it's cold here." Having just returned from nearby Nanisivik, where the weather had apparently been much better, he claimed that "anywhere is better than Resolute Bay. This is what happens when hell freezes over."

However, contemporary colonial ideologies were deployed in the starkest terms by wealthy, transient traffic through Resolute airport. Adventure tourists often manifested profoundly socially superior, even potentially racist, attitudes towards the people of Resolute. Consider, for example, the following passage from an account of journeying to Ellesmere Island by Canadian explorer and travel writer Jerry Kobalenko:

> My partners in crime likewise commented on how they felt like "stray cats" or "bag ladies" in Resolute. Sometimes we [visitors] came prepared to spend substantial amounts of money and still felt like second-class citizens. Resolute's icy-heart had seen too much. Too many foreign polar bear hunters willing to drop $20,000 in four days, leaving thousand-dollar tips in their wake. Too many North Pole expeditions with half-million-dollar budgets. Too many planeloads of doctors and stockbrokers. Too many cruise ships disgorging ladies in furs (mink, not caribou) and men with glaring, corporate eyes who plunk down two thousand dollars for a narwhal tusk during their one hour on shore. Resolute is like Las Vegas: impossible to impress, no matter how much money you throw away ... But, at worst, Resolute could be so disheartening and humiliating that I would return home never wanting to go north again. Yet, like cold, fatigue, soft snow, high winds, and partners from hell, I soon forgot Resolute and remembered only Ellesmere. (2002, 5–6)

Resolute, for Kobalenko, precisely because it is a social space, is disheartening. It acts only as logistical fulcrum necessary to facilitate passage to the imaginary and *uninhabited* landscapes of northern Ellesmere Island.

Notwithstanding some important interactions, then, it would appear that the historic socio-spatial segregation of the scientific base and the local community and associated ideologies persist, or are even reinforced, in contemporary Resolute. However, this lack of interaction is completely subverted on a specific day each summer.

5.2 | Vehicle parade on Canada Day passing Polar Continental Shelf Project base complex, July 2002

Enacting the Ritual of Canada Day

Each year, on 1 July, Canada Day is celebrated throughout Canada. There are geographical variations in the sincerity with which these festivities are held across the country. Significant pageants are generally associated with certain parts of Canada, such as Ontario, Manitoba, and British Columbia. Similar to many civic festivals in Western nations, however, there is often reluctance to take such projects seriously. Even in Vancouver, for example, the marking of Canada Day is often treated with an embarrassed disdain. National celebrations do not fit well with the imagined Canadian sense of self. However, in Resolute, Canada Day is one of the very few occasions when the scientific base community and the local Inuit community interact. And what is most interesting about the mainly Inuit hamlet of Resolute is the degree to which the community participates in symbolic acts of *Canadian* national identification.

In Resolute, then, there is an annual Canada Day parade. This is followed by a series of celebratory events encompassing a large bonfire, a communal barbecue, and a series of competitive river-crossing races. The Canada Day festival is organized by the hamlet council. As well as local dignitaries, such as the mayor of Resolute, the participants in these events include almost every member of the community. In these circumstances, the PCSP scientists signify the federal presence in Resolute. In short, the scientists *embody* the Canadian state. I want to argue that these events can be seen as part of a carnivalesque ritual that involves the relief of community tensions.

Canada Day commences with a lunchtime parade in which the fire truck, festooned with local children, leads a convoy of all of the vehicles in the village up to the airport complex (figure 5.2). The fire truck is driven by Alistair, the Inuk who is employed to drive the water truck for PCSP. At the airport, more vehicles join the convoy, before all participants travel a couple of kilometres across the tundra to a nearby riverbank for the other festivities.

A group of scientists wanted to join the parade in July 2002. All of these individuals were Canadian, coming from regions across the country, except for myself as a British ethnographer.[13] The request to participate was important, because PCSP scientists perform a crucial role in relieving tensions by becoming figures of ridicule during the carnivalesque festivities. As this narrative of events should suggest, from the very first instance, the scientists are unwitting actors in an unfolding local comedy.

The scientists believed that the parade was to start in the centre of the hamlet at 1 p.m. Mainly because they had, perhaps purposefully, been misinformed of the parade's starting time, as they travelled over to the hamlet to join the convoy, the scientists met the fire truck and other vehicles coming in the opposite direction up the single-track road towards the airport. The vehicles had apparently left the community at 12:30. The scientists therefore had to reverse their vehicle, in front of the entire parade, for over six kilometres, until a wider embankment allowed space for the truck and other vehicles to pass. This involved the complete convoy of vehicles travelling past and waving to the scientists, whilst also tooting their car horns and mocking them, before the scientists' vehicle was able to join the back of the parade. The parade then continued towards the airport and base complex, where all the vehicles paused for about five minutes. Once at the base complex, most of the eager scientists, except for James, a biologist who felt that he was too old for such activity, climbed onto the roof of the fire truck, together with the Inuit children.

It is important to note that, except for Alistair in the truck cab, the only adults travelling on the fire truck were scientists. While we were on top of the truck, a female graduate student asked an Inuk child if he would get cold without a hat. Astutely, he replied, "No, we won't, but you might!," causing all twenty or so children to laugh hysterically. The scientists were thus also mocked by the children. However, their exotic status as outsiders within this *local* celebration of national identity meant that the scientists tended to play up their role at these cross-cultural moments. On the roof of the fire truck, scientists sought to entertain the children by showing them their expensive digital video cameras. Expensive personal goods acted as significant novelties in these spaces. Indeed, it could even be argued that new technological artifacts helped to stabilize the contact ritual.

When the convoy reached the river, a large bonfire constructed the previous day was set alight. Most of the scientists then spent their time pruning sticks with penknives, creating toasting forks for the children to roast marshmallows on the bonfire. Again, little attempt was made to speak or interact with adult Inuit.

Whilst food for the community was prepared on a large barbecue, races were organized from one bank of the river to the other (figure 5.3). These races involved young children, older children and adolescents, and adult men and women. Many of the younger scientists participated fully in these activities. In doing so, they realized that, like local power-holders such as the mayor and RCMP officers, they should be seen to take part but not to be victorious. For example, a number of scientists theatrically fell into the river while leading their races. This further enabled socially acceptable mockery of the federal state embodiments, thereby facilitating the carnivalesque dissolution of usual hostilities.

There were also signs of social tensions *within* the village being dissipated. This was especially evident when the river race for women came down to a very close finish between the Inuk wife of the (non-Inuk) mayor, and Emma, a former Co-op worker and PCSP employee (figure 5.4). It was obvious that the community wanted what was perceived to

5.3 | Community celebrations on Canada Day, near Resolute, July 2002

be an "Inuk" victory. The connection to employment by the Canadian state therefore appeared to be less divisive than that of being married to a non-Inuk. After Emma's victory, the majority of attendees at the barbecue made a point of congratulating her individually.

Despite interacting with children in a paternalistic way, and participating through the inversion of power roles in the river-crossing races, scientists also attempted to initiate conversations with some of the Qallunaat inhabitants of Qausuittuq. Both scientists and these local people would bemoan military activities in the area. Often perceived as leaving polluting traces in the landscape, neither scientists nor other local non-Inuit were happy with the supposedly unsupervised activities of the Canadian Forces in the Arctic. There was a more powerful form of federal presence on Cornwallis Island about which these groups shared antagonism.

The lack of conversational interaction between scientists and adult Inuit notwithstanding, the general atmosphere of the Canada Day performances was seemingly very friendly. However, James the biologist

5.4 | Community river-crossing races, near Resolute, July 2002

was obviously uncomfortable and left the festivities early. It was evident that he was not enjoying watching the activities and did not want to participate in the races. James had undertaken environmental science on the territories of other indigenous groups in southern Canada. Despite his discomfort, he commented, "The atmosphere is not as unfriendly as an Indian reserve."

Inuit Nationalism in Resolute

There was a general sense that the Inuit of Resolute wanted to convey, at least ostensibly, pride in some sort of Canadian identity. The Canadian flag was prominently displayed from the fire truck during the entire duration of the parade and other festivities (figure 5.5). Every child had little Canadian flags painted on both cheeks. In many ways, there was much more celebration of symbolic *Canadianness* than I ever observed during two Canada Day events that I participated in whilst living in Vancouver.

5.5 | Canadian flag displayed during Canada Day celebrations, July 2002

Moreover, the question continually proffered to scientists by Inuit on Canada Day was "Are you Canadian?" If the scientists answered in the affirmative, they were encouraged to join in with the festivities or to share in the food. The implication was very much that, on this particular day, both Inuit and Qallunaat should remember their commonalities rather than their differences.

The carnivalesque moment of Canada Day allows Inuit to stress Canadian aspects of their identity. In doing so, such practices problematize facile binaries between agents of the imperial state and indigenous subjects in the colonial present. Furthermore, apparent nationalistic celebrations by Inuit help undermine, perhaps ironically, simplistic understandings of patriotism in multicultural states.

The performance of Canada Day in Resolute similarly raises important questions about Canadian nationalism and its connection to Arctic science. The supposed need to defend Canadian sovereignty is often still deployed in arguments for increases in federal research funding by northern scientists (England, Dyke, and Henry 1998; England 2000, 2010). The ethnographic evidence indicates that, notwithstanding their Inuit identity, in Resolute people still wish to celebrate Canadianness. As the director of the Nunavut Research Institute (NRI), Bruce Rigby, put it, "I think you'll find in the North, northerners in general are probably much more nationalistic than anyone in southern Canada."[14] In response to this question of nationalism, the executive director of NRI, Mary Ellen Thomas, states,

> This debate came up recently when there were claims of *who owns* the North Pole. And that the Russians had made some claim to the ownership of the Pole, and the Danes had made some claims to the ownership of the Pole, and the Americans had made some claims. And the question was, "Well, why isn't Canada out there, claiming the North Pole?" And my answer to that was, "Because we don't need to." Because the reality is we're already here, we have a presence here and an identity here. And this whole question of sovereignty to me is just a *joke*. I mean, it serves someone else's interests. But for the people who *live* here, the Inuit people, this is *their land*.[15]

Given the creation of Nunavut, Inuit become able to celebrate all forms of identity, including that of "Canadian," in whichever way they see fit. As Rigby puts it. "I think you'd find an awful lot of people who are just a little bit uncomfortable with someone else's nationalist road. *It doesn't make them any less Canadian*. Or any less nationalistic. But it just means that people aren't willing to allow themselves to play into somebody else's agenda. And I think that's one thing that I've learned over the years here. People are very strong up here."[16] That the demonstration of such Canadian identity also allows the relief of external and internal social tensions, as Turner would have argued, is to be expected of any such carnivalesque event.

Conclusion

Ethnographic investigation of the performance of Canada Day in Qausuittuq allows us to think more seriously about the notion of the

Arctic as Inuit homeland and the implications for debates about national identities. Resolute is absolutely not a wasteland of nobodies. It is home to people who have suffered interference and indifference from the Canadian state. These people have continued to maintain culturally meaningful lives in the face of such dominating institutions, and in doing so create a rich, social world.

Like Bakhtin's carnival, the celebrations of Canada Day outlined here are not to be understood as simply a loyalist reinvigoration of the state. Rather, Canada Day involves the "temporary suspension, both ideal and real, of hierarchical rank" in Resolute (Bakhtin 1984, 10). In doing so, the celebrations facilitate a "special type of communication impossible in everyday life" between scientists and inhabitants of the community (10). Moreover, Inuit relations with the Canadian state are characterized by ambivalence – an ambivalence made manifest in the rituals enacted on the first day of July.

Social life in the Arctic involves the use of moments such as Canada Day to deal with the continual embodied presence of the federal state. Inuit lives are constructed through events that allow the alleviation of tensions internal to the construction of the social order. The account outlined thus supports Jean Briggs's classic ethnography of the methods Inuit use to resolve conflict (Briggs 1970; Stuckenberger 2005).

Moreover, such events constantly enact the colonial present. Research in critical Inuit studies has stressed the complicated relationships between aboriginality and Canadian identities in northern Canada (Stern 2006a, 2006b; Stevenson 2006). For Inuit in Resolute, the birth of Nunavut has not placed them *outside* the Canadian state but has instead allowed the reformulation, during carnival and ritual, of cultural relationships within it. The creation of Nunavut appears to have resulted in complicated refractions through Inuit identities, which require more attention from students of nationalism.

Resolute is an itinerant home to the scientists, mechanics, pilots, and other maintenance staff who enact scientific practices in the Arctic. It is through encountering such agents that Inuit experience the everyday Canadian state. As this chapter has tried to indicate, the interactions between these groups, whilst historically involving dispossession and racism, have important ramifications into the present. Only when the Arctic is envisaged as a social space, encompassing places wherein different people interact, resist, and celebrate, is a full understanding of such regions facilitated. It is through ethnographic observation of such practices that the assumptions of many social and political theories of the nation and the colonial encounter are revealed. Inuit self-determination has unleashed the potential for reformulation of national spaces through the enaction of quotidian life.

Although always structured by power, the complexity of these human interactions needs to be understood for the future subsistence of such places in the high latitudes. As the Arctic regions become ever more crucial to global environmental futures, it is imperative that these interdependent histories of encounter be remembered and that their diverse voices are documented. In outlining some of this history, this chapter has argued that the mundane geographies of the Canadian Arctic have very important consequences. It is less that indigenous resistance reverses colonial axes of power and more that self-determination allows revelation of the complex intertwining of cross-cultural understandings.

6 Emotional Practices and Play: The Quotidian Provenance of Logistics

> The learning of an emotional vocabulary is one of the essential skills of an ethnographer. To survive as a competent social being, the fieldworker must learn how to interpret, if not actually feel, the finer shades of anger, pity, or whatever the host population specializes in.
>
> Andrew Beatty, 2005a, "Emotions in the Field," 23

Emotional Geographies and Affectual Anthropologies

Just as social analysts have neglected the constitutive processes of scientific authority, so too have they often ignored the emotional constituents of science. Similarly, although geographers have drawn attention to the importance of the epistemological and ontological dimensions of scientific practices, rarely have their respective emotional remits been considered. Discussions amongst geographers of science have therefore begun to call for heightened attention to the potentialities of ethnographic methods (Powell 2007a). Through the deployment of such sensitivities, the emotional practices involved in doing science might begin to be revealed. At the same time, geographers of science have rarely engaged with the vibrant discussions occurring in geographies of emotion and affect. In what follows, I attempt to meet these challenges through an emotional ethnography of environmental science in the Canadian Arctic. In doing so, this chapter attempts to derive, through ethnographic practice, some intimations for discussions in emotional geography.

Emotional geographies are concerned with the spatial mediation and articulation of emotions (Anderson and Smith 2001; Bondi, Davidson, and Smith 2005). This has resulted in focus upon the spaces through which emotional geographies are enacted. Such work has directed attention to the emotional intersubjectivity of social interaction and specifically to "issues of relationality which are so profoundly embedded in our everyday emotional lives" (Thien 2005, 453). For Thien, if emotions are relational consequences of human interaction, then their embedding must relate to other positionings within the social order. Debates about the relationality of emotions have therefore become oppositional to other work that has assumed the primordial status of affectual response (McCormack 2006; Anderson and Harrison 2006). This geographical research on affect has been criticized for ignoring the important power-geometries that affect human lives (Thien 2005; Tolia-Kelly 2006). As Tolia-Kelly argues, it "is thus critical to think plurally about the capacities for affecting and being affected, and for this theorization to engage with the notion that various individual capacities are differently forged, restrained, trained and embodied" (2006, 216).

I should stress that rehearsing and adjudicating between these different epistemic standpoints is not my concern here. Rather, I want to suggest that these debates amongst geographers have tended to operate in parallel to those that have taken place across anthropology and sociology during recent years. In this chapter, then, I draw from these related literatures to help revisit the current impasse amongst geographers of emotion and affect.

Put starkly, relations between emotions and modes of social organization have been central to the sociological tradition – "to talk of social structure *is* to talk of emotions" (Beatty 2005b, 55; my emphasis). In what follows, I argue that geographers of emotion could learn from debates about the ethnography of emotions, most evident in work by anthropologists.[1] As Kay Milton puts it, "Anthropologists are professional observers of human situations, and emotions are present in everything they observe" (2005a, 215). Deriving from ethnographic experience, Milton argues that rather than being solely social in origin, "emotions are essentially ecological phenomena" (2005b, 25). In so doing, Milton outlines an argument that emotions are mechanisms via which humans learn *through* their environments (Milton 2005a, 2005b). Emotions are central to this learning process. Similarly, Svašek (2005) draws attention to the emotional intersubjectivity of fieldwork.

The lesson to be drawn from the anthropology of emotions, I want to argue, is the central importance of the practice of fieldwork. The very experience of fieldwork, or "being there," involves a series of emotional encounters (Tonkin 2005; Josephides 2005).[2] During fieldwork in Java, for example, the ethnographer is "lacking an explicit set of rules or taboos, [and] one must, as it were, *feel* one's way" (Beatty 2005b, 64; original emphasis). As such, developing awareness of emotional codes is part of the process of *becoming* in the field for any successful ethnography. For Beatty, it becomes critical to maintain "heightened attention to what happens in the field" (2005a, 18). This requires the emplacement of ethnographic fieldwork at the very centre of the geography of emotional practices. By focusing on the "varying pragmatic contexts" of social life, ethnographers can make a central contribution to the understanding of emotions in it (34). As Beatty argues, this "would entail a shift away from worrying about affect, and from inner/outer quandaries, to what I have called *emotional practices*. It would also mean a more exact reporting of what pass as emotional episodes in the field, for from such material we weave our theories" (34; my emphases).

After Beatty, then, I use *emotional practices* to indicate those displays and namings of social emotions that can be grasped only through their recording in ethnographic context.

Playing with Scientific Practices

This chapter contributes to debates on emotional practices by revisiting a tradition of Arctic ethnography – that of *play*. However, it does so by widening the communities of practice involved in Arctic social life to include various emotionally competent beings, such as environmental scientists, bush pilots, and, especially here, base support staff. This theoretical mission is accomplished by examining the ethical and epistemological consequences of observing what I term *spaces of play*. Through ethnographic research at a scientific base in the Canadian

High Arctic, I argue that the spaces that compose such an organization are contested and thus involve competing but always interacting communities of practice. Within these communities, social interaction is undertaken between field scientists and logistical support staff, or between indigenous and non-indigenous groups. This has consequences for the individuals involved, not least the development of an emotional register of reactions to an imposed hierarchy of importance for scientific and logistical practices. Moreover, these reactions result in dissent being *performed* in and through spaces of play. Whether the emotional geography of such spaces *should be recorded at all*, then, becomes a guiding question of this chapter. In order to understand the implications of these statements, it is crucial to grasp the role of play in the Arctic ethnographic tradition.

David Riches has argued that Inuit studies have been hampered by overemphasis on the documentation of tradition, at the expense of embedding discussions within theoretical debates in social and cultural anthropology (1990). However, this argument about the supposed atheoretical stature of Inuit studies does not hold in anthropological discourses of emotion and affect. Indeed, there has been a long tradition in Arctic ethnography of the anthropology of emotions, best represented in the work of Jean Briggs on the "patterning of emotion" in Inuit family life in her path-breaking *Never in Anger* (1970, 7). As Briggs showed through sensitive explorations, emotional interaction affects the ethnographer as much as the agents under investigation. This provides critical insight for geographies of emotion. As Briggs puts it, "Cross-cultural studies of emotion have so proliferated in recent years that to list a few authors is to offend many. Relatively few of these, however, have focused on the *socialization* of affect" (1998, 267). Understanding the social embedding of affect, then, requires ethnographic experience.

Moreover, Briggs developed an emotional anthropology through another classic trope of Arctic ethnography – that of "play." Briggs argues for "the emotional power of the games" within Inuit society (1987, 13). It is through learning from these games that Inuit come to maintain feelings and values with respect to aspects of social life. In her evocative ethnography of Chubby Maata, a three-year-old Inuit girl, Briggs (1998) illustrates the importance of playful games for the development of emotional motives behind the experience of everyday life.[3] And as Nicole Stuckenberger demonstrates in her excellent ethnography of Qikiqtarjuat, play is still central to theorizations of Inuit sociality (Stuckenberger 2005). For Stuckenberger, within the seasonal morphology of Inuit social life,[4] games carry a critical ritual significance. During religious festivals, such as "Christmas,"[5] participation in games allows structured competition and cooperation and thus the energetic development of social relationships. As Stuckenberger concludes her study, "Thus the notion of play, so fruitfully developed by Jean Briggs, acquires a special meaning: it constitutes a modality which allows the community to connect itself to transcendental agencies, and to establish a sense of community and cooperation between the participants. All involved are well aware, though, that these relations are fragile and can only be fully realized in play" (213). Play, then, is a central trope in accounts of social life in the Arctic.

My study of field scientific practices revolves around the performance of the everyday at the PCSP base and respective fly-camps. Jean Lave argues that the everyday does *not* "denote a division between domestic life and work" (1988, 14). The everyday is instead formed from what people do in quotidian cycles of activity. Such conceptualizations are important in ethnographies of field science, where demarcations between domesticity and labour are particularly difficult. Moreover, recent work in anthropology has begun to take notions of "play" seriously (Gable 2002; Lury 2003). In what follows, any casual distinction between work and play is problematized by the spatially contiguous nature of domesticity, science, and indigenous practices at an Arctic field station. This is developed through the deployment of the notion of "spaces of play," that is spaces where resistance and dissent are performed *through* games and mimicry. Social interactions, as I have argued, are critical in the constitution of Arctic field science. These practices occur as much in spaces of play as in workspaces. And it is precisely for this reason that, in this chapter, I will focus here on "facilitators of science" rather than "scientists."

By concentrating on the everyday lives, opinions, and values of the base staff at PCSP Resolute, I contribute towards anthropologies of labour that take the role of work in the life projects of human agents seriously (Corsín Jiménez 2003). The seasonality of labour practices in the High Arctic is such that these *spaces of play* tend to emerge infrequently and in reaction to local variances in environmental and social conditions, such as malign weather or personnel changes on base. This can mean that such occasional spaces emerge during the frenetic routines of the base and, whilst ephemeral, are also peculiarly carnivalesque.

Dislocated Labour: Being a "Newfie" in Resolute

For any study of the conduct of scientific practice in Arctic Canada, therefore, it is important to note that indigenous peoples are not the only marginal voices in the region. The most obvious of these are the base technical staff from Newfoundland, introduced in chapter 4.

There was constant casual disparaging of the important PCSP employees from Newfoundland. The first thing base manager Yvonne told me, when taking me on a brief tour of the warehouse after I arrived at PCSP, was, "All the staff are from Newfoundland and they all speak *Newfoundese*, so you might find them a little difficult to understand. We do."[6] Edward, when explaining how he felt safe only in a Twin Otter (as opposed to any other aircraft), because he knows how to fly them, joked, "Though I would be afraid here, because all the mechanics are Newfies!"[7]

Indeed, it seemed that the staff from Newfoundland could be discriminated against even more that Inuit. During one lunch-time conversation, for example, a scientist told a story about an Igloolik school, where the pupils had very bad completion rates. The scientist believed it was because the community kept flying up head teachers from Newfoundland, rather than making internal promotions from amongst the local staff.

These sorts of jokes were told with relative affection, and cannot be theorized as *solely* articulations of racism. Moreover, the Newfoundlanders were equally mocking of other forms of difference encountered around Resolute. An RCMP officer who lived with her female partner in Resolute came in for particular attention from the Newfoundlanders. As Barry, the stores manager from Newfoundland, put it bluntly, "Two women living together in Resolute Bay, when the ratio of men is twenty-five to one! I know it's bad up here, but I didn't know it was that bad!"[8]

The Newfoundlanders are probably the most critical element in the functioning of the PCSP. These employees work exceptionally long hours. When the 2001 sealift came in a few days ahead of schedule on 24 August 2001, the Newfoundlanders worked through the night to unload it. They were then on duty again less than four hours later at 7 a.m.

Although it was initially difficult to be accepted by these members of the PCSP hierarchy, as I was initially associated with other communities of practice, such as the scientists or the base management,[9] over the course of fieldwork, it became easy to talk to the Newfoundlanders during space-times of play. At such moments, in response to circumstances provided by weather conditions, or by personnel changes on base, such as when staff were arriving at or departing from PCSP Resolute, these spaces emerged. Towards the end of a field season, for example, a space of play appeared because staff believed that "the season was over." Another occasion was that, although operations on the base had a further eight weeks to run, a senior member of PCSP management was arriving from southern Canada the next morning, so it was a final opportunity to celebrate without surveillance. Such occurrences usually involved alcohol, music, games, and discussion, which included gossip about scientists on base or senior members of the PCSP hierarchy.[10] These spaces of play are therefore doubly carnivalesque, both because they were areas for jollification and they involved the occasion to subvert authority (Stallybrass and White 1986).

Edward, as base manager, frequently stressed the importance of the staff from Newfoundland. It was very difficult to plan logistics, complained Edward, when all the PCSP seasonal staff returned to Newfoundland for the winter. The staff were effectively unemployed for most of the year, and their embodied expertise would be useful for planning logistical arrangements. As Edward put it, these members of staff possess crucial knowledge and experience, but the costs involved to bring them to Ottawa or to convene some other sort of meeting during the winter was prohibitive.[11] As Edward argues, "We are such a small organization it is difficult to do things … Polar Shelf works because of the people, the 'Newfies' and the cooks, up here, but they are getting older, and they are getting pissed off as they get older. They don't get paid enough for being up here and away from their families, and they basically get abused by the government. We have been fortunate because we have had 'Newfies' and there are no jobs in Newfoundland … But we can't afford to lose the experience."[12]

Most senior scientists also recognized the importance of the Newfoundlanders. Trevor Bennet, from a government research institute in Manitoba, argued, "PCSP is an unsung hero … But it is the guys on the ground who make PCSP. These guys spot the naive, new scientists and show them how to live in the field. Without these guys, beakers would die."[13] Another scientist, Quentin Lefebvre, argued that morale among the Newfoundlanders was "well down compared with the eighties. These people make Resolute run. Polar Shelf could run with anyone at the camp manager level – all they have to do is decide where a plane should go. But Polar Shelf could not run without the mechanics and cooks."[14]

The reason for these constant reminders of the importance of this labour in the PCSP from both managers and scientists was that the Newfoundlanders perceive themselves as marginalized from the institution. This is, in many ways, a consequence of the management practices adopted by PCSP.

Towards the end of the 2001 season, for example, Edward was pleased that the sealift arrived ahead of schedule and had been unloaded within a few hours. For Edward, this meant that the base staff should be able to finish the season early, and thus go home, so he told Barry, "We should get a reward for getting everything done quickly." "What," Barry replied, "two boots in the arsehole?"

There appear to be two main sets of consequences of the seasonal labour provided by the Newfoundlanders. The first is the relations with the hamlet. In short, the Newfoundlanders decrease the amount of employment available to Inuit. As Edward puts it, "Way back, Tuk was busier than Resolute, then Resolute grew in importance. We used to hire 'locals' on seasonal contracts, like Rachel [the cleaner], and Alistair who does the water truck and other odd jobs. We used to hire more, but when Tuk was shut down, the [Resolute] 'locals' lost out, because the Newfies came across to work here."[15]

This problem is exacerbated by the actions of a man who holds public office in Resolute. Although married to an Inuuk, he is not local. He has commercial interests in the area that he runs in competition with those of the institutions of local government, such as a private hotel in the village to compete with the Co-op Hotel. He is also ostentatious in displays of his wealth relative to other residents.[16] This is important, because his businesses tend not to hire locally, but to bring in employees from Newfoundland, which has had consequences for local elections as well as the Resolute wage economy.

The Newfoundlanders employed by PCSP are thus only part of a larger migratory labour force from the province that arrives in Resolute every summer. Moreover, the PCSP Newfoundlanders are also employed, by this same man, on an occasional "cash-in-hand" basis in the evenings after they have finished shifts on base. The dubious nature of his employment practices is not lost on the Newfoundlanders. On one such evening, for example, a group of scientists asked Craig, who dealt with tents and equipment for the field, why the PCSP Newfoundlanders had been moving gravel for this entrepreneur. Craig joked in response, "I can't tell ya'. Or if I tell ya', I'd have to kill ya.'"

The employment of Newfoundlanders therefore influences all relations between Inuit and southern Canadians in Resolute, because of the development of competing forms of proletarian identities. In many ways, this is similar to the growth of a French-Canadian manual labour force in Iqaluit in the 1970s (Brody 1975).

The second consequence of the reliance on seasonal labour from Newfoundland is the *personal* costs of working for PCSP. It is difficult for anyone to cope with long periods of isolation, and these feelings of alienation were most evident at space-times of play. A simple example is Barry's comment: "We get good food, but it is the same every day each week. It gets you down, but it's not the cook's fault." Researchers have produced psychological studies of cognitive stressors and coping strategies involved in long-term residence at polar bases (Barbarito, Baldanza, and Peri 2001; Mocellin 1988; Mocellin and Suedfeld 1991). However, such studies cannot quantify the impact of events that occur *away* from the bases, for those with whom base staff have personal relationships, often because of these absences. During the first season that Gary, a Newfoundlander responsible for building maintenance, was in the Arctic, he spent fourteen consecutive weeks on the PCSP Ice Island.[17] Arriving on the Arctic Ocean during the uncomfortable environment of early February, conditions were made more difficult because "that was the first time I had ever been away from my wife." A conversation between the Newfoundlanders, during a space of play, illustrates this point:

GARY: It takes a certain kind of *person* to work up here. You have to be able to handle the isolation. It was really bad on the Ice Island.

BARRY: It takes a certain kind of *wife*.

Both Edward and Barry, as Gary put it, "are on their second wives." Barry's child was six months old before he saw her for the first time.

Another contribution to low morale is staff uncertainty about how long they will be kept on base. They are often "sent south" early, cutting short their envisaged season of employment, without warning. When a senior manager returned to PCSP Resolute in 2001 towards the end of the season, he had decided that he would send another manager, two Newfoundlanders, and a cook, home on the next flight south. Only the senior manager and two Newfoundlanders would remain until the base was closed down for winter.[18] This was upsetting news for some staff. These seasonal contracts effectively form the annual salary for the staff, so such logistical decisions that are made as PCSP suffers from budgetary pressures have massive ramifications for the Newfoundlanders and their families.[19]

All the Newfoundlanders thus feel isolated from the decision-making of PCSP.[20] The marginalization of the Newfoundlanders was therefore expressed through frequent, implicit challenges to the authority of management. These acts of resistance were heightened during spaces of play. Gary made jokes about the perceived lack of technical know-how of the senior management. Further, Barry argued, "We're the ones who make the real decisions. We never listen to Andrew when we're loading the planes. I just want to know where the group are heading and for how long."[21] On another occasion, when angry over a particular decision, Barry was more forceful: "We do the fucking work. In the office, half of what they tell you is bullshit."[22]

At other times, Barry tended to be more circumspect: "I'm happy ... The management always seem to be increasing their numbers over the years. Whereas we, the ones who do the work, have always stayed about the same. Though it's the same with any job, I suppose."[23] Barry had a longer-term perspective than most staff or scientists. As he argues, in many ways PCSP Resolute never changes: "I have seen it been as bad as this twice before. It goes in cycles."[24] At the same time, I always had the impression that the Newfoundlanders were *proud* of the occupation of their lifetimes. As Barry noted, "I could've wrote a book from what I've seen. I've been to the North Pole twice. Not many men can say that. My job is to help the science and the scientists. They go out in the rain and the snow and live in a tent, and we decide how much fuel they need. Ninety per cent of what scientists want is wrong."[25]

There are similar consequences for the seasonal kitchen staff. By being in the base office one afternoon, I inadvertently witnessed an outburst by a member of staff who had just found out that she was about to be sent home. Although she appeared nervous and anxious that I was there, she launched into her understandable tirade anyway.[26] Kitchen staff always stressed how much they enjoyed their job, and they viewed PCSP Resolute as *home* because they spent so much time (usually around six months of the year), with the same people, every year.

It was through learning to participate in these spaces of play, therefore, that I began to find evidence of hidden stories and agendas within the PCSP. However, these spaces of play often involved "loose talk," and this leads to anxieties for the analyst about how to represent such observations.

The Ethics of Learning from Play

Participation in spaces of play, I argue, results in epistemological and ethical quandaries. Such spaces emerge from resistance to the official narratives of scientific practice in the Canadian Arctic. It is from the observation of activities in such spaces, therefore, that fuller descriptions of scientific and logistical practices in Arctic field science are possible. In short, material derived from learning in spaces of play gives better access to the communities of practice involved in constituting PCSP Resolute.

However, precisely because such spaces resist management practices and official narratives, they are usually inaccessible, and perhaps should always remain so. Spaces of play become accessible only by participating in play. And this results in ethical anxiety for the ethnographer. Agents appeared to be much more relaxed and informal with an ethnographer who was participating in a space of play. Indeed, some of the spaces that I have partially documented here are, as the result of local regulations regarding alcohol, problematic. How is previously negotiated "informed consent" complicated during spaces of play? How should material derived from such spaces be represented? The ethnographer must negotiate between fidelity to representation of practices and articulation of often repressed voices and experiences, whilst anticipating the consequences of representing spaces of play for informants long after return from the field. In this discussion I have been careful to censor much discussion, as well as making the usual attempts to disguise identities. But are such efforts sufficient?[27]

The presentation of this material has therefore required careful attention to ethical issues, because it has been difficult to represent voices and emotions without making evident the identity of the articulating individual. I should note that some of these opinions may be surprising, even disturbing, to those expecting an unqualified account of scientific practice at PCSP. Needless to say, my recording some of them here does not mean that I fully subscribe to any or all of them. However, this effort at careful representation has been crucial to making my argument and to fulfilling my ethical obligations as an ethnographer of practices. This raises an emotional dimension of the practice of ethnography that is all too often neglected – that of *guilt*.

Participating in spaces of play has resulted in *guilt*, that most powerful emotion, about my incomplete attempts to represent some of the results of my learning. This, of course, is not an uncommon reaction for emotional geographers. Rebekah Widdowfield outlines a common set of responses, experienced during fieldwork with lone parents in the West End of Newcastle, such as anger, distress, and ultimately demoralization. The resulting sense of powerlessness, as she puts it, led her "to question both my ability and desire to carry on with the research project" (Widdowfield 2000, 205). In an essay on the emotional responses of the researcher to the field, Liz Bondi illustrates the commonality of feelings of guilt and shame at the inadequacies inherent in the research process: "I felt guilty because, despite my efforts to be honest, people were spending time with me and were telling me all about their experiences, perhaps at least in part, on the basis of a misapprehension about my capacity to offer anything in return" (Bondi 2005, 238).

In an account of fieldwork in Papua New Guinea, anthropologist Lisette Josephides argues that the "strongest motivator which emerges from the ethnographic vignettes is the feeling of *resentment*" (2005, 72; my emphasis). Through this formulation, Josephides argues that resentment for the ethnographer "is the feeling of not being acknowledged" (86). Rather than resenting the communities of practice encountered in the field, I feel that it is the notion of being unable to represent spaces of play sufficiently that causes anxiety for this ethnography.

For Bondi, through the process of qualitative research, the practitioner must *suspend* decisions about feelings so as "to reflect on emotions in their full richness and complexity" (2005, 241). It is not that I am describing shame for poor research design or incoherent thought about research ethics; rather, I argue that guilt is a necessary emotion during such research and must therefore be thought through and resisted, lest it become methodologically disenabling. It is within these discussions that emotional ethnographies should take central place.

Conclusion

This account has stressed the importance of ethnographic experience to the study of emotions in social life. As fieldwork involves a set of affectual encounters between the ethnographer and other practitioners within competing and complementary communities of practice, doing such work can contribute to debates about emotional geographies. The corporeal rigours involved in ethnography, the need to understand, interpret, and empathize with emotional codes, sensitize the analyst to allow sensible theoretical discourse regarding emotions. By outlining some of the emotional processes of scientific research in the Arctic, I have argued for the broadening of these intellectual encounters in geographies of science and emotional geographies.

In attempting to convey the viscerality of everyday scientific practices, then, I wish to contribute to the development of ethnographies *from the body* to develop a "carnal sociology," as Loïc Wacquant (2005) puts it. Attempting to study the emotional remit of scientific fieldwork uncovers aspects that are often best *revealed* through analyses of spaces of play. But as the consequences of these descriptions are uncertain, I am still anxious all these years later about whether I should have participated, and thus learnt from them, at all.

7 Hidden Voices? Competing Visions and the Everyday Governance of Arctic Science

Every moment that I'm here, I feel I'm very fortunate. I'm one of the very, very tiny fraction of people that can come to a place like this and actually work here.

Malcolm Ramsay, *Vets in the Wild*, 2000[1]

Introduction

In this chapter, I attempt to help revivify the study of field science by recording some of the hidden voices concerning the future of PCSP. I have argued that it is emotional impoverishment to fail to understand the empathetic relationships that develop between the researcher and research subjects during observation of practices in the field. It is only through developing empathy with the positioning of the subjects involved that it is possible to understand fully the institutional spaces of PCSP. In doing so, the researcher develops an ethical responsibility to record these relationships accurately. It is possible for a theorist to fail to understand the importance of these relationships only if she has never taken seriously the difficult work of actually describing practices.

Moreover, there is also an analytical advantage in presenting some of these hidden voices. As the institution of PCSP is constructed by the practices of so many different individuals, this chapter helps guard against the assimilation of difference within a study of quotidian scientific activities in the High Arctic. It is my stress on the importance of *interpretive* ethnography that facilitates my resistance to universalizing tendencies in some observations of practice. This also reminds us that maintaining "analytical distance" is a constant challenge in the composition of any ethnography, and thus helps us to be reflexive about unarticulated agendas within any description of practice (Herbert 2001, 309).

The following chapter thus devotes greater presence to *competing voices* across every level of the PCSP, and the impacts of the responsibility for the representation of them upon the ethnographer. In order to make this case graphic at the outset, I want to draw from a reflexive moment in my fieldwork that emerged when I returned to Resolute in June 2002, and specifically the tragedy associated with the BBC TV program *Vets in the Wild* filmed at PCSP Resolute in 2000.

Vets in the Wild and the Importance of Interpretive Ethnography

At my first meal back at Resolute, I encountered some hostility from Catherine, a graduate student. When it became apparent in conversation, for example, that I had helped in the kitchen the previous season, she tersely stated that this was no longer allowed because of a new waiver system. As I was describing my project to other scientists at the table, I initially believed that Catherine was antagonistic because she thought of herself as *the* local PCSP historian, as she had also helped out in the kitchen and talked to scientists and base staff.[2] However, it was only when she mentioned a story about some British journalists filming scientists at Resolute, and the Newfoundlander staff wanting to name the film "Beakers on Ice," that things become clearer. And when she stated that Polar Shelf had had two fatalities in its recent past, I began to comprehend something of the sources of this enmity.

When Catherine had last been at PCSP Resolute, during the 2000 field season, a group of BBC film-makers, from the program *Vets in the Wild: Polar Bear Special*, had been in Resolute. Not only had this group been British; they had, by all accounts, behaved arrogantly around the base and village.[3] Tragically, the two scientists who had been filmed as they studied polar bear behaviour, Professor Malcolm Ramsay, of the University of Saskatchewan, and Dr Stuart Innes, from the Department of Fisheries and Oceans, died in a helicopter crash very soon after the journalists had departed south. This was an understandably contentious incident, but I mention it because, as an outside observer from Britain, I was inevitably viewed in the context of this tragedy.

In many ways, I think this event skewed my entire fieldwork at Resolute, as well as my initial attempts to gain access to PCSP support and to conduct interviews with scientists long before I began any empirical research. Although I never mentioned this to anyone directly, I gathered clues from oblique references. As I have noted, some people were especially wary of me when I first arrived in the field in 2001. Many scientists and staff in Resolute wanted to know why a "Brit" was interested in the Polar Continental Shelf Project. I believe that I was initially guilty by association of nationality with these journalists, and by a similar "potential meddler" status.[4] The situation also required even greater sensitivity in the ethnographic observation of scientific practices, because my sense was that scientists were more anxious than they might normally have been.

However, none of this contextualization could proceed through *just* describing practices. In what follows, then, I make the case for the interpretive presentation of stories involving individuals across the organization of PCSP. In so doing, I not only argue for the heterogeneous composition of an organization involved in field science, I also show how this network of PCSP is inhabited by sentient, reflexive people. I want to indicate further the importance of the labours and personal consequences of those who facilitate and create field science in the Arctic. Having observed many of these social practices, I believe that they are often best expressed through oral discourse with participants.

The Consequences of Arctic Flying

As discussed in chapter 4, pilots had relatively high status within the social hierarchy on base. However, there were still personal repercussions for the pilots in working on contract for PCSP. One evening during sked, Peter, a pilot, came into the office. He was due to fly south in two days. He had flown for Polar Shelf in the past, but now worked for First Air and so was no longer on contract. He was obviously demoralized and was talking of a helicopter crash the previous year, in which people had died. The helicopter engineer had just finished working on the machine before it crashed. Peter would never fly helicopters because "it is a matter of when rather than if" there is an incident, and Twin Otters are also dangerous. For Peter, there was a current problem with the Arctic expertise of pilots, as the good pilots were retiring and the younger ones would no longer take the necessary risks. Peter argued that he does not "tell my wife half the shit that goes on up here. It is not like it was. They will have to call it a day if the pilots stop, because the beakers won't be able to do their research."

For Peter, it was critical for scientific research that the pilots take risks. Peter was therefore planning to stop flying in the Arctic and get a job in the south with a firm with a FedEx contract. This would be "easier flying, better money and closer to home. Only an hour and a half back to Fredericton after work."

This was very interesting in that pilots contracted by PCSP seemed to be losing a sense of national duty as was common among early pilots, such as Bert Burry discussed in chapter 2. Although Burry was contracted, there was such a sense of excitement and of duty to the *nation* that risks were automatically taken. Another helicopter pilot was increasingly upset, for example, towards the end of the 2001 season, because he wanted to get home for the long weekend but had to wait for the scientists he was transporting to call an end to their fieldwork.

Although some of Peter's comments must be seen in the context of First Air losing the PCSP contract to Kenn Borek, the impact of the isolation on such people is massive. Stuart, another pilot, had a pregnant wife when he came up in the 2000 season. She had an emergency, induced labour of twins at six months. Both children died, and the mother was in intensive care and nearly died as well. This happened on a Monday, but Stuart was unable to get south until the next jet on the Wednesday.

Peter was not optimistic about the future, because when the mine at Nanisivik closed in 2002, he believed that First Air would stop the jet service, and though there may be something in the summer when the load was heavy, "we can't fill them with tourism." Peter was very interested that I was studying Polar Shelf, as if this was another indicator of the decline of the institution. As he put it, "Canada aren't [*sic*] putting enough money into our own backyard."

Being "the Boss" at Resolute in Times of Temperance

The next level of the institution to consider are the base managers. My presence was generally accepted by the managers. However, the managers were perhaps the most uncomfortable group of individuals when I was observing them. In one incident that illustrates this situation, a dangerous chemical, peracetic acid, was found in one of the hangar "cages" of the Bylot Island field party. The jar containing the chemical had apparently been in the storage area for years, but hidden at the back. The managers were discussing the need to dispose of it, but were not sure how to do so safely. Charles, a junior manager, stated, "We need a chemist up here," and joking to Edward, theatrically turned to me, and asked, "Are you writing that down?"

However, the two managers with whom I conversed most often were also the more senior. As these managers were responsible for drafting and implementing the rules governing behaviour on the base, I relay some of these dialogues in the following sections. This discussion will show how the base managers attempted to maintain a national vision of PCSP, within Canadian Arctic science, whilst also dealing with many of the personal implications.

Andrew appeared to be generally pessimistic about the future of northern science in Canada: "Only the US could do things on their own up here, and even they aren't that bothered. Politics works by taking money from one pot and giving it to another. There's never any new money. And every decision is made by consideration of how it looks nationally, what will get more votes, rather than internationally. But climate change needs big science, run by lots of countries together. And new technologies are changing the way we do things."[5]

At the same time, he also had a vision of the historical evolution of PCSP through the different decades of its existence. These transitions provided some reasons for optimism for the future of PCSP: "Polar Shelf has changed through every decade. In the seventies, they were still doing science at Polar Shelf. And Resolute was the busiest airport in the world when there was oil and gas interest. Then, in the eighties, there was a transition to logistics. By the nineties, almost 50 per cent of the logistics was for universities, plus we had the growth of clients from the territorial governments. What will Polar Shelf be in this next decade? We still don't know. But every decade has seen a transition."[6]

In many ways, like his employees, Andrew himself felt isolated from the decision-making regarding PCSP. Andrew did not become overly anxious about this, though, as he put it, "Most of the politics of Polar Shelf has worked behind closed doors. Like the initial Polar Shelf. It could have been about continental shelf science, but it could also have been about defence and sovereignty."[7]

Most important of all, there was also the sense that Andrew felt that PCSP, and thus his life's work, had not been sufficiently recognized by Canadians. However, he still maintained a sober wit about this: "Polar Shelf has never had too much public interest, but logistics never do. We have supported movies and documentaries by National Geographic and the Discovery Channel. But who cares? No one watches the credits at the end of the movie. Who gives a shit who Grip 2 is?"[8]

This lack of recognition of the activities of PCSP appeared to be sometimes replicated by other federal departments, such as Indian and Northern Affairs Canada. In July 2002, for example, an environmental inspection team, comprising a representative from Indian and Northern Affairs and two RCMP officers came to PCSP Resolute to speak with Andrew. The inspector, by his own admission, had been in the job for only four months, and obviously had little knowledge about northern Canada. The inspection team was in charge of an environmental cleanup operation, and had been flown around the Arctic, investigating old scientific and landing sites for abandoned waste and fuel drums. During the conversation, Walder stressed that the database, composed by PCSP, of all the landing strips used by all federal agencies in the High Arctic over the last forty years, together with existing fuel quantities and empty drums, would be useful. The existence of the database was unknown to the inspection team. Andrew suggested that, in the future, he would be able to email digital photographs of relevant sites to the inspection team before they "come up here," as this "will save the government money in the long-run." This may be an isolated example of a novice inspector, but it illustrates that there can sometimes be difficulties of co-ordination between the branches of the federal government operating in the Arctic.[9]

It is important to reiterate, in this context, that PCSP is the main federal government operation in Resolute, if not the High Arctic, and often performs many crucial duties without getting sufficient credit. In April 2002, for example, PCSP provided logistical support for the

Kigliqaqvik Rangers expedition to the Magnetic North Pole.[10] A nineteen-year-old Inuk was lost near Pond Inlet during my fieldwork, and one of the PCSP pilots looked for him, whilst flying another project nearby. The RCMP in Pond Inlet also inquired about borrowing a PCSP helicopter for this purpose.

Frustrated Visions: Performing the Governance of Canada

A theme that emerged strongly in my fieldwork was the apparent neglect of the roles played by logistics personnel in the mapping of Canada. This is evident, for example, in published accounts of postwar cartographic methods in Canada (Thomson 1999). Of all the PCSP staff on base, it was Edward Freeman who felt most disaffected about a perceived lack of recognition of the work by his staff.

A Twin Otter pilot, Edward was employed by private sector logistics organizations in Resolute before he was approached by PCSP for a vacant base manager post. Edward's career has involved work across a number of PCSP research bases: at Resolute, then the Ice Island, then Tuk for about four years, and then a return to Resolute. When Edward started at PCSP, there were five base managers, at Eureka, the Ice Island, Resolute, Tuk, and a "free manager" who moved around. In the early and mid-1990s, the field season was longer, and during the summer busy period (which was then mid-June until the end of August), Edward would be on duty for twelve to eighteen hours per day.[11]

From conversations I understood that, as only Resolute was operating at the time, and base managers now had less freedom to make judgments for their particular base, they thus felt assimilated into a stricter decision-making hierarchy.[12] I got the impression therefore that the shrinkage of PCSP had resulted not only in a loss of jobs, but also in a decline in job satisfaction for the remaining staff. For Edward, having experienced the private sector, this was a consequence of working for a federal institution:

> With the private sector, if you worked hard you got rewarded, but if you screwed up, you were fired. With the government, if you work hard you won't be rewarded, but at least it is very hard to fire you … At Resolute, the staff work hard because they know the scientists so well. It is certainly not for any financial reward.

> The younger staff here are not so bothered about the money, but they want to make a difference. The government will have to loosen the reins, otherwise they'll walk, because there are opportunities out there for them. My generation, we couldn't do that. We needed the money.[13]

The constraints of employment within government meant that Edward appeared to feel that his ideas about logistical innovations were stifled. During another conversation, Edward mentioned that he would like to be able to learn more about logistics from Antarctic science. There had been an exchange at the higher levels, but he wished to see exchanges at the base camp and logistics level. This would likely require funding, but he thought it might also be possible via email correspondence. Isabel Mitchell, for example, wanted a particular snowmobile on the Devon ice cap early in the 2001 season, but PCSP were not able to load it aboard a Twin Otter. However, Isabel was able to tell Edward how they did this in Antarctica and it was quite simple. Similarly, in 2001 Kenn Borek had the PCSP contract for the first time, and their pilots tended to have much more Antarctic experience than Bradley's.[14] This suggested potential changes in protocol to Edward.

Following up on this theme, I asked Edward whether current procedures, such as the PCSP Arctic-Antarctic Exchange Program, could include base staff. He replied,

> I have suggested it, but … The problem is that within the Government of Canada everyone protects their own regimes, and there are also the problems of co-operating with foreign governments. Just in the [Canadian] High Arctic we have the Department of National Defence, the Rangers, Environment Canada, and the Coastguard. If even we could get them together. But we would need the political will and people willing to mash it out. We need to sort out Canada before we could get other countries to co-operate. We are always having conferences on the Arctic, but at the high level, not at the logistics level.[15]

If PCSP were to receive a bigger budget, Edward's first priority would be *safety*. He would ensure that there was sufficient staff and that a better GPS aircraft tracking system was in place. When he was base manager at Tuk, Edward set up a system using GPS to monitor every PCSP aircraft by modem. He wanted a similar GPS system to be set up through Resolute, but thought there were, at the time, too many financial constraints. Moreover, under the current system, PCSP need two managers in the office whenever it was "busy," and that was why Edward wanted a flight watch system, to help reduce this burden. According to Edward, convincing people of the value of such a system was difficult, but it was getting easier as there had been an expansion of GPS usage across federal departments. However, Edward thought that another possible solution would be to write a GPS requirement for all aircraft into PCSP flying contracts. This might increase the contract rates, he argued, but insurance costs for PCSP would come down. As Edward put it, such a system "would not stop accidents happening, but it would allow us [PCSP] to get there before someone bled to death."

Edward also wanted to reopen PCSP Tuk, "but only if there was sufficient demand."[16] Another of Edward's thwarted ideas was that PCSP should supply satellite telephones that could also send emails. Then, an email could be sent in when camps had no traffic for radio schedule and also so that PCSP could suggest a calling time for each camp, rather than them all calling in at once, immediately after sked, and creating a risk of the base staff missing something important.

These stories may appear initially as simply the expression of the frustrations of middle-management common in any organization. However, employment in the PCSP was different precisely because of the personal consequences and the national vision with which it had always been associated. The consequences of working away from home for many months of the year meant that it was a notion of national service that employees like Edward used to justify their occupation. Edward, like many of the base staff, had had problems during his career in maintaining personal relationships and close friendships in the south. As he noted, "I have more friends up here than at home, and they all come back every year."

Given the ramifications of working for PCSP, Edward wanted a re-articulation of a vision for PCSP: "There is such low morale on the base. We put in long hours away from our families, and we expect respect and the knowledge that we're doing a good job. Sure everyone always wants more money, but that isn't crucial. Is this the end of Polar Shelf? I don't know honestly. There is no future for the young people involved, so they are doing stuff on the side. Other people are looking for other jobs. The money isn't good, but we don't mind because we love the job. It has been my life. I'm looking to retire."[17]

At another moment, Edward protested, "It has to get busier up here … I have invested a lot of my life in Polar Shelf and I want it to be good again."[18]

The problem for many of the PCSP base staff was that they felt that an organization to which they had committed their lives, at a level far beyond any traditional notion of labour due to the long periods of absence from their families, was in terminal decline just as they approached retirement. Reflecting on the future of PCSP, Edward commented,

> I can see them closing it down. It has been dying a slow death for the past six or seven years. Maybe it's not that slow. In the next two or three years something has to change. Either we'll be closed down, or we'll get more money. But any change won't come from inside Polar Shelf, it will come from the scientists or the government. Canada let its military shrink, and now we are throwing loads of dollars at it.[19] I can see that happening to Polar Shelf. They let it shut, and then throw money at it, and then in five or ten years it will be reinvented. That would be a shame because there is so much here. But things change. The Weather Stations and DEW-Line stations are all gone. But too much would be lost if they let Polar Shelf shut … I could see Polar Shelf being taken over by private companies, or by Nunavut. But then they would just hire us and pay us less.[20]

However, what was most demoralizing for Edward and his staff was paucity of public knowledge and recognition of the activities of PCSP:

> In any Canadian city, people don't know about PCSP. Most scientists don't know about PCSP, and even some Arctic scientists haven't heard of PCSP. PCSP needs to raise its profile. The top management of Polar Shelf should get more publicity for us. They should go to universities, scientists, and foreign agencies and get the applications in to justify their role to the government. The image of Polar Shelf that exists is good, but it is not well known enough … There were recent discussions about

> changing the name and logo of PCSP, but they were stopped. It took us forty years to develop this name, so if we change it we go back to "ground zero."[21]

The importance of a national vision to be achieved through scientific practice in the Arctic, which was discussed in chapter 1, was therefore still critical to Edward's sense of job satisfaction and personal achievement. Articulations of Canadian sovereignty over the Arctic surfaced in everyday discussions at Resolute. For example, a Canadian Coastguard officer, visiting the base after the 2001 sealift, suggested to Edward that there were US scientists conducting research in the archipelago who did not recognize the islands as Canadian. Both the officer and Edward bemoaned the fact that there was never Canadian money behind anything. According to the Coastguard officer, "the Swedes and the Germans" even wanted to take a Canadian icebreaker, *Louis St Laurent,* to Antarctica. As the Coastguard officer reflected, "The Germans are invading and we don't know it."

This concern about sovereignty was therefore inextricably linked with notions of the national project of scientific practice in the Arctic. If only, Edward continued, the Canadian public could comprehend the importance of Arctic science:

> The Canadian public is not bothered about Polar Shelf or Arctic research. But a scientist we supported found out about how to store food at an optimal temperature to make it last longer. This helped lower the cost of food in the south. If only people could see the resource potential and economic benefits, then they would be in favour of greater support for northern scientists. But people don't know about northern science because Canadians don't brag about it, unlike the Americans. If we could get the science into a form that could be disseminated, then people would understand why we have to support northern science.[22]

Edward was pragmatic about this. He understood that it was difficult to make the importance of High Arctic science intelligible to the public,[23] and even more difficult to justify logistics: "Science is easy to cut because it's difficult to justify. Logistics are not that difficult to cut. Even worse, they are concrete, in the sense that you can say a Twin

[Otter] costs this, fuel costs that, food costs that. We knew the scientists and got close to them, and we could see it was important. But the costs of science are intangible. Science has to prove itself to Joe Blow Canadian."[24]

The current recognition of PCSP in Nunavut was also relatively low, according to Edward. Although communities such as Resolute, Grise Fiord, Arctic Bay, and Pond Inlet had a good knowledge of the organization, there was little awareness among Inuit in Iqaluit on southern Baffin Island, far from the main area of PCSP operations.[25]

However, for Edward, the creation of Nunavut still offered an opportunity for PCSP. Although Edward appeared to view Nunavut as a resource hinterland, there was a sense in which PCSP could be revitalized through the expansion of the territorial economy: "Demands for energy will increase because the population in the south is growing. Nunavut is rich in resources, the native population is concerned about the environment and wildlife. But those people will have to be employed and we will need the energy. It is just a matter of time before the exploitation of Nunavut's resources becomes economical ... If the federal government put more dollars into Arctic science, it pleases Canadian scientists and the international community. And if the economy improves then there should be more money for science."[26]

Misogynist Spaces? Being Female at PCSP Resolute

Having discussed some of the hidden voices of labour and base managers at Resolute, and argued that they still carry many of the discursive formations of science in northern Canada as a national project, I now move on to discuss some different everyday exclusions at PCSP. These are issues of potential gender discrimination.

There are many tales of gender discrimination in field science, and historical events at PCSP Resolute are no different. Nicola, as an older female scientist, relayed an instance of her being prevented access to certain field sites at Prince Patrick Island, because there was only one bathroom at the base camp. Similarly, she had not been allowed into Resolute at all in the early years of her career, as women were allowed only as the spouses of scientists.

In the ethnographic present, younger female base staff appeared to suffer minor forms of discrimination. They were often teased, jocularly but significantly, by base managers and pilots about decisions and comments.[27] One female base employee, as she was single, was often called "a lesbian," even if convivially. This supports the arguments of sociologist Arlie Hochschild (1983), about the ways in which gendered practices are folded within *emotional labour*.

There were starker examples. Some pilots, when communicating by radio with the base whilst airborne, attempted to speak solely to *male* managers to discuss landing conditions and weather forecasts. On one occasion, during a radio conversation between a female base manager and a female airborne helicopter pilot, a number of other pilots and engineers also happened to be gathered in the office. All these men were mocking the caution of the female pilot, as she was checking the location of emergency fuel caches. The assembled personnel did not believe that she would in any way need to know where these caches were. The base manager was much more understanding, and as she explained to the airborne pilot, a male pilot in the office quipped, "You women stick together." It was apparent that other pilots and engineers, all being male, thought this female pilot was "way too soft," and other comments were passed on this issue on numerous other occasions. Such incidents involved public displays of chauvinism and misogyny by many of the pilots and aircraft engineers.

Another important dimension of gender in field appeared when "domestic space" intruded into discussions. If too much evidence was presented by individuals about personal relationships, such as partners or children, then they could begin to be thought of as "not being good field persons" and, in turn, poor scientists. Female scientists most often underwent this diminution of status, and might have suffered gender discrimination as a consequence. When gathered together and usually if there was a mixture of research groups, they often discussed issues such as marriage and personal relationships, and were not taken seriously as scientists by other research groups as a result.

In many ways, these forms of discrimination, at least by the base staff, appeared to be enacted in order to facilitate the functioning of the base. For example, incoming telephone calls from the south, that were broadcast over the intercom system on base, were almost always for female graduate students to the exclusion of all other social groups on base. A student got a telephone call from her mother saying that she was "worried about her being up north." On another occasion, another female graduate got a telephone call in the office, and Andrew complained to Edward about the "nancy-girls getting calls" from boyfriends that were coming too late at night and preventing him sleeping. Edward agreed, and moaned, "They aren't that bothered anyway. I tracked one down and she said, 'It's OK – they'll call back'!"

This was inconvenient for staff, because there were only two telephone lines, which were routed through the base office and bedrooms of base managers, in case of emergency. Moreover, they could be an operational impediment if they occupied a line when a camp was attempting to reach PCSP with an urgent message or emergency by satellite telephone. Interestingly, despite large numbers of calls for female graduate students, both late at night when staff were sleeping and during busy office hours, I never observed a single occasion of a male scientist getting an incoming call. This does not mean that such communication did not happen, of course. Rather, it means that male scientists were making outgoing calls, and doing so more covertly, if at all. It would be impossible to conclude whether this was a reflection of the way spouses and relations in the south constructed younger female scientists, or some sort of behaviour of the female scientists at PCSP themselves.

As a result of this complicated relationship between pragmatism and discrimination, even as a participant observer striving to maintain detachment, I found myself sometimes making everyday judgments about who was a "good field person." It was difficult to find too much sympathy for some undergraduate field assistants who, having arrived at Resolute, refused to join their parties at field sites, arguing that they missed their husbands. A young lecturer did not want to fly out to a field site because they usually had problems with polar bears and the weather in that area, so parties were often delayed.[28] She was due to fly home for her three-year-old's birthday, and as all the family would be there and she missed her partner, she did not want to be stranded at the field site.[29] The most striking performance of not being a good field person occurred when a sombre male scientist talked very personally about his wife and how much he missed her, over dinner, to a group of incredulous scientists. Having revealed publicly that he believed that "she is too flat-chested and has too much weight on her arse," the scientist then began to assess the quickest route and flight itinerary that he could take home.

However, one of the most fascinating intrusions of domestic space into PCSP Resolute was the announcement that two students had become engaged to be married during the 2002 field season. This appeared to add an uncomfortable dimension to relations around the base. They were able to share a room, despite the supposed gender segregation of areas of the "beaker" accommodation. Increasingly amorous behaviour marked their activities around the base. As a number of people commented, this was uncomfortable for other scientists at the base, but was also seen as unfair for the staff who were all away from families and friends for long periods.

Articulating Practices: Scientists Talking Science

A further set of hidden voices were also articulated by scientists in discussions among themselves at PCSP Resolute. As distinguished Canadian earth scientist Alan Merton joked, "Ninety per cent of Arctic science is telling stories and ten per cent is science."[30] Scientists often gathered in the kitchen at around 8 p.m., and discussed the politics of research and issues of the day, whilst eating extra desserts. Such conversations reveal concern for the practice of field science in the Canadian Arctic.

These discussions, often involving individuals from across different research groups, involved complaints about the price of flights and the inconvenience of getting to PCSP Resolute with First Air. As one put it, "We'll all have to move to Ottawa or Edmonton!"

A number of the themes of discussion are summarized here, taking particular care not to provide information that could identify individual scientists. It was decided that, for this section, the dating of comments from field notes would allow the identification of individuals, because of the timing of their usual field season, and thus when they are in PCSP Resolute. Therefore, in the following sections, all referenced conversations are from field notes recorded at some time in August 2001 or June–August 2002.

A basic theme of debate involved the repercussions of the creation of Nunavut. Although scientists always fully supported the new territory, there was always the underlying opinion that there was a relationship between its creation and the lack of funding for northern science. Although most saw basic merit in the Government of Nunavut's principle of decentralization, for one scientist there were significant cost implications for communications and transport. As this scientist put it rhetorically, "What could scientists from the south do with those helicopter hours?"[31]

A related discussion topic was the licensing process involving the Nunavut Research Institute (NRI) in Iqaluit. There was general distrust of the NRI staff and a sense that the institute was wielding too much power over research. A scientist expressed a commonly held view when she quipped, "Blink and you've missed it" about NRI.[32] She believed that many in the research community think that problems over research licences emerge from Inuit grievances over land claims, "but it isn't them at all, it is people on power trips." The implication, for this scientist, was that it was the Qallunaat employees of specific agencies of a newly powerful territorial government who were causing problems for them.

Another scientist was even more outspoken about scientific licensing procedures:

> Licensing takes away the flexibility of northern research, which is crucial because of the weather. Last year [2000], [name withheld] and I had to leave early because of bad weather. We had wanted to shift to a site fifty kilometres down the coast, but we couldn't because we knew we would be crucified because we didn't have a licence for that site. The whole licensing procedure annoys me. I get a letter back from some bureaucrat, who knows nothing, saying, "Yes, this looks interesting," and they're not saying it in a good way, it is a patronizing way. It's all "big fish in a small pond." I know it's interesting, NSERC have accepted that in giving me a grant.

The licensing situation could be genuinely difficult for scientists. Any scientific research on Melville Island, for example, required licences both from the Government of Nunavut and the Government of the Northwest Territories. However, many senior scientists were somewhat cavalier in their attitudes to licences. Although all licences and permits were supposed to be with PCSP three weeks before arrival of the first member of the field party at Resolute, on a number of occasions I witnessed base managers reminding scientists to give them copies as they were boarding flights to their fly-camps.

The Funding of PCSP and Flying Hours

The issue of PCSP funding was constantly debated by scientists. A subject of passionate discussion was the allocation of "flying hours." Researchers argued that they always built "circling time" for flights into their requests for PCSP support. However, these extra hours were usually pruned when offers of support were made by PCSP. Circling time was crucial for scientists, such as when landing at a new site, or perhaps with an inexperienced pilot. It was critical to have pilots who were able to land Twin Otters at rough camps, not just on prepared strips. The accounting of hours then became critical, because a scientist might think, for example, that he had used twenty hours, but PCSP might issue a bill for, say, thirty-two hours at the end of the season. According to various scientists, it was never apparent how these totals were reckoned. It was very difficult to find out without asking a base manager directly, which in turn could be construed adversely and thus have implications for future applications.[33] And yet even an extra three hours of flying could cause a project to go $4000 over budget.

It also appeared that scientists were unsure how the cost-sharing procedures of PCSP worked. "What happens," one asked, "if a pilot arrives and tells the scientist in the field that he can only take two boxes because he has had to pick up some empty fuel drums in an environmental cleanup? How many hours does that cost the research party?"

In another conversation, a group of scientists were complaining about a helicopter pilot for being unhelpful, being unprepared to wait for his passengers in the field, and not telling them his plans. As one of the group put it, "He thinks he's just transport. He doesn't see the science." Many scientists complained if they felt that they had a pilot who was uncooperative, because they believed they were at the mercy of the pilots' decisions.

Scientists often became frustrated when they felt that they were not getting the flying time they deserved, whether because of unsuccessful applications or not having access to aircraft when in the field. In 2002, the current topic of anxiety related to PCSP funding was a rumour about plans to begin charging Canadian university projects at full cost recovery rates, like non-Canadian projects were. However, a senior scientist warned, "You have to be careful of crying wolf unless it is a real wolf. But politicians are slow and can only remember one thing. That's why we have to stop cost recovery for Canadian universities."

Along with a group of Canadian Arctic scientists, however, this individual had been lobbying federal ministers to increase PCSP funding and to keep the name "Polar Continental Shelf Project."[34] But there was a general feeling among scientists that they did not understand the decision-making processes of PCSP. Another scientist summarized this point: "A question that a lot of us ponder is whether PCSP can implement a policy on its own, or whether it is simply an organization that implements other policies. Where is it stated that PCSP cannot support research in the Western Arctic? Or how much is allocated to supporting university research? Are these just off-the-cuff decisions with massive implications?" At the same time, this scientist articulated the common anxiety of almost every PCSP client whom I interviewed or spoke to more informally: "I don't want to slag PCSP. I owe my career to them."

Inuit Participation and Philosophies of Field Science

Scientists sometimes discussed how to get "locals" involved in science. A university professor, for example, suggested that scientists needed to get Inuit involved in small ways in their proposals, such as providing one night's accommodation on the way into the field, and then that person became a collaborator, the idea being that by gathering enough Inuit collaborators, the federal government would be unable to ignore the funding application.[35]

University and government scientists always stressed the importance of PCSP to field research. A government researcher who used remote sensing techniques, for example, stated that he could still "ground-truth" out of communities without PCSP, but the work "would be scattered and unsystematic, so the science would suffer. Without PCSP, my research would just be a series of case studies."[36]

Moreover, for government scientists, gaining PCSP support validated the importance for that field research to be undertaken by the host department. But through residence at PCSP Resolute, this same government researcher had made contacts and formed new collaborative research ideas: "But the bean-counters don't see that. The spinoffs are unexpected though and become invisible … The value of an institution is sometimes overlooked. The people at Polar Shelf don't want to lose the name [of PCSP] because of the 'brand name.' So what if PCSP doesn't work on the continental shelf anymore? Everyone knows what it is, so what does it matter whether they work on the continental shelf?" For this researcher, remote sensing would not replace people having to go into the field, to "ground-truth," but it would increasingly help identify key areas for field research to be undertaken.

These conversations therefore often encompassed interesting questions from the philosophy of field science. Many appeared to believe that the philosophy and practice of Arctic science were being radically transformed, as an older *observational* field science was being ceded to Inuit. In one such discussion, two university professors agreed that there was no future in wildlife surveys for graduate students, because this research could increasingly be performed by "locals," as there was little difficult training involved.

For a government researcher, Canada was ignoring the need for fieldwork across the biological and geological sciences. Increasingly, he argued,

> research grants are being given for modelling or the analysis of historical data, but not for the collection of new data. Publication just takes too long when it is based on fieldwork. No one does taxonomy anymore. I need lichen identification for my "ground-truthing," but there are now only two people left in Canada who can do this, and one is fifty-seven years old! There has also been a decline in [environmental] monitoring, because this requires long-term fieldwork, and you can't get the money over the long term because of the length of time required for the payoff. We North Americans have very little sense of the importance of the past. People in my department just throw out old documents and data.

Over lunch, Yvonne, the base manager, discussed the success of the field season with a professor from the earth sciences. This scientist, although an experienced fieldworker, had never worked in the

Arctic and stated that he "was not prepared for the biting cold of Ellef Ringnes [Island]." Yvonne and the scientists then developed an interesting debate about Arctic science:

> YVONNE: I only knew what I was doing halfway through my MSc. When I was writing up, I thought, "Right, I have to come up with a hypothesis."
> SCIENTIST: That seems to be common in any sort of field science.
> YVONNE: Yes, but it is exaggerated up here.

Yvonne then gave an example. Whilst she was undertaking her fieldwork for her MSc, an Arctic fox chewed through the lead to her automated "data logger" at her field site at Expedition Fiord. Yvonne needed a soldering iron. However, soldering iron is classed as "dangerous goods" by carrier aircraft and she was not able to get one. Eventually she found a helicopter engineer who had a soldering iron, but she lost weeks of data. Similarly, at Eureka in 2002, a research student was unable to do any fieldwork for over a week, because there were polar bear sightings in the vicinity of her field sites.

The conversation continued:

> SCIENTIST: There seem to be all sorts of unique logistics problems when doing research up here. The amount of effort required to get one data site up here is orders of magnitude higher than that required down south.
> YVONNE: Yes. People doing research up here seem to just go "Data, data, data," thinking they can sort everything out down south. I was jealous of fellow grads who had a nice site, one hour's drive away from Uni, which they could go to once per week to download their data. Some even installed modems so they could have real-time data.

Discussions about the unique difficulties of Arctic research reverberated around all sorts of spaces on base. As well as the scientists, Andrew Walder told a story in the office of one PhD student who needed just one further season's worth of data for his dissertation. However, for the past three years, fog and polar bear scares had prevented him from accessing his field sites. As Andrew put it, "I had to talk to the PhD committee and explain that it wasn't his fault. They thought the kid was dumb!" Yvonne, listening to this, said, "Use what ya' got and change your topic."

Following conversations about the peculiar factors of Arctic field science, I gave Yvonne a copy of the paper by Fred Roots (1969) on logistics (discussed in chapter 2). For Yvonne, the unique factors of practising Arctic science include the isolation, and thus distance from help, and the safety considerations, such as keeping within a safe distance from a radio when at field sites in case of accident or polar bear sightings, and the difficulties that "Arctic vets" can face when trying to reintegrate "down south."[37]

A day or so later I went to talk to Yvonne about the Roots paper during a quiet evening in the office. Yvonne agreed with Roots to some extent, such as that developments in logistics technology were continually making things easier, and that a certain type of person or scientist is attracted to do work in the Arctic. However, "I take issue with him claiming that these logistics problems make science less rigorous in the Arctic than down south. There are world class scientists up here. Sometimes things go wrong, but people do world class science. It just takes longer."[38]

During another discussion, Michael Plummer, a distinguished hydrologist, argued that the permafrost specialist J.R. Mackay became so respected as a field scientist in Canada precisely because he conducted *field experiments*. For Plummer, "Geomorphology, in the British tradition, was all about casting an 'eye' over the landscape and not about processes."[39] It was critical, in the 1960s, for Canadian geomorphology to begin to study *processes*, because there were, according to Plummer, "too many brilliant explorations of topology by Mark Melton, or slope asymmetry by Barbara Kennedy that did not show processes."

Younger scientists also articulated this distinction. A graduate student drew a demarcation between the practices of "measurement" and "sampling" in field science: "The sampling I have been doing with [name withheld] at Eureka is more rush, rush on good days, and then taking the bad days off. But the measurement and process studies I did on Cornwallis at [name withheld]'s sites involves a more regular routine."

My impression therefore is that using an old school definition of geomorphology and historical geography as a basis for a critique of fieldwork, as Gillian Rose does, is problematic because that view of fieldwork has been discredited in geomorphology since at least the 1960s and the enthusiasm for process studies and field experiments.[40]

Often, after conducting interviews with individuals, I found my interviewees being quizzed in the kitchen by two natural science graduate students. If I ever stayed in the kitchen in such situations, the conversation stopped as they had obviously been discussing my research and my aptitude to undertake it. However, occasionally these conversations were more enlightening than the answers given by the scientists in a more formal interview. One university professor, after talking to these graduate students, noted, "Maybe [in the interview] I didn't stress the camaraderie of PCSP enough. I still think PCSP *is* logistics though."

The next morning, after the interview with this professor, I met him in the bathroom.[41] He wanted to know how many interviews I had conducted at that point, and what ministers and deputy ministers thought about PCSP. He also wondered whether scientists were saying the same things as the bureaucrats. This professor stressed that he had not emphasized enough that "there has to be a realistic alternative if Polar Shelf changes. Canada always expects money to come from somewhere else, rather than stepping up to the plate. Will you produce something that will have an impact? Then we can interview you."

Social Spaces of Scientific Exclusion

The final set of hidden voices that I will discuss involves the spaces of exclusion *within* the community of field scientists at PCSP Resolute. In both interviews and conversations, many scientists mentioned the supposed camaraderie of the community at Resolute, and the scientific interaction facilitated by the base. However, I *observed* only a little of this co-operation and intellectual exchange. For example, a scientific party at Truro Island had used a paraglider in order to survey sea ice, and discussing that over lunch, Alan Merton was very interested in it as a possible technique by which research groups could cover more ground in the field.

Female graduate students often stressed the importance of PCSP Resolute as a place where discussions over dinner allowed development of new research projects, or simply the lending of equipment.

Interestingly, it was comprehensively but exclusively female graduate students who made this argument. And as they always seemed to be drawn into discussion by older male scientists or pilots, to the exclusion of male graduate students, one wonders whether they might have been slightly naive in their optimism.

A number of scientists stressed that the introduction, over past the four years, of satellite TV in the beaker building had impeded intellectual interaction among the scientists. However, other scientists argued to the contrary that it had facilitated it, because it drew a number of scientists with spare time to a communal area.

At the same time, it was evident that there were certain cliques, resulting from old friendships and connections from graduate school. There were also obvious understated rivalries between scientists in similar fields. Certain Canadian scientists appeared unpopular, especially those who formed part of multinational research groups, perhaps because they were associated with the decisions of the PCSP Steering Committee and Scientific Screening Committee.

As noted in chapter 4, the scientists who particularly suffered in the status hierarchy, and thus in priority for flying and explanation of the situation, were usually from less prestigious Canadian universities, and those who did not have a history of PCSP support.[42] Such researchers were often ignored, for example, when they came into the office to check whether they were flying that day, whilst more established researchers were dealt with first. However, all these social rules were completely unspoken, and could be understood only after prolonged observation. One such researcher went on to cause a series of difficulties in an Inuit community, simply by not staying in the local hotel. Although this researcher was inexperienced in Arctic fieldwork and did not appear to be a "good field person" from the start, I believe that some of these problems could have been prevented if he had been more accepted by the scientific community initially, and thus advised on how to adapt to the peculiarities of northern research.

One researcher and his undergraduate field assistant found PCSP Resolute to be an unfriendly and humourless place. The researcher made the very interesting point that I was one of only two researchers to make an overture to them during their four weeks on the base. Most scientists were accepting if he made the effort to approach them, he thought, but some blatantly ignored all attempts. He argued, "The only people who are friendly to others [at PCSP Resolute] are those without their own group." When I put to them the common argument that Resolute was a site of scientific ideas emerging from interaction, the professor replied that it was "bullshit." The field assistant went further, joking that "Resolute is a place of stagnation, where ideas die." Observing groups' frequent excitement when they were able to leave Resolute to get to the field, the undergraduate suggested that PCSP "was like a retirement home," because the inhabitants were overly excited when they could go on a trip.

Much of this cynicism appeared to result from an unsuccessful field season, and I also had the sense that the researchers believed that they would not be allowed back to PCSP Resolute in the future. Although their responses must be examined with substantial caution, these excluded researchers did make a serious point. A group of seven Japanese scientists were beginning an interdisciplinary project on Ellesmere Island in 2002. They had excellent funding and wanted to make connections with Canadian scientists but were not able to. This was not because they did not speak good English, as most were fluent, but more because they were never really approached by any Canadian scientists.

In summary, then, Resolute facilitates exchange between many scientific programs, but it is very hard to get interaction with research groups *who are viewed as somehow different.*[43]

Conclusion

By studying the fieldwork conducted under the auspices of the PCSP in 2001 and 2002, I have argued that field practices both around the base and at field sites were structured by rules policed by the base managers. In doing so, I have attempted to show that it is not possible, as a researcher, to describe these practices without beginning to relate to the individuals who perform them. Observing quotidian practices results in the development of emotional commitments to the participants of research, which results in an ethical responsibility for the accurate representation of such articulations. Moreover, it is impossible to understand the functioning of Arctic field science without taking such an expanded view of practices.

In this chapter, I have thus depicted some of the constituent elements of PCSP, as it is wrong to circumscribe the importance of these individual activities for human lives. At every level of the PCSP, for Inuit and Newfoundland labour, for pilots and base managers, and for different types of scientist, there are frustrated voices. Although complicated and competing, they hold in common that PCSP, through its support of scientific fieldwork in the High Arctic, is part of a vision of a national project. Like many national projects, some are excluded from this vision. In this case, it is usually those who are not construed as competent in the field. By attending to all these voices, though, I have attempted to breathe life into accounts of scientific practices, as well as providing greater analytical power to fieldwork stories.

EPILOGUE: REQUIEM FOR A *CANADIAN* ARCTIC?

All ethnographic writing is constrained by the temporality of the ethnographic present. The field cultures described here have, however, persisted into the ethnographic future. In summer 2007, in the wake of the Russian scientific mission to the North Pole, many of the sites discussed in this book returned to widespread international attention. What was missed in the global media circus was the scientific justification behind the geoscientific mapping of the submarine Lomonosov Ridge (Powell 2008b, 2010). The institutions established by the UN Convention on the Law of the Sea, 1982, such as the Commission on the Limits of the Continental Shelf, have resulted in a specific juridical-scientific settlement that requires huge investment in the geosciences in the Arctic.

The Conservative prime minster of Canada, Stephen Harper, had secured only a minority government in February 2006, and pursued a number of projects to solidify national support. In November 2007, in his opening speech to the new session of the Canadian Parliament, Harper outlined a new Northern Strategy. He argued that Canada needed to be more active in the Arctic. Harper was fully aware of the echoes of his Conservative predecessor John Diefenbaker when outlining his new vision for the Canadian Arctic. Harper's felicitous mantra, "Use it or lose it," came to dominate Canadian Arctic policy discourse during recent years.

Following the lead of the other Arctic littoral states, through 2007 and 2008, Canada announced a series of measures, including a new fleet of icebreakers, greater military presence, and a deep sea port, that were envisioned to re-establish Canadian sovereignty in the High Arctic. Furthermore, the Canadian government announced a re-vitalised program of field science centred on a new, technologically innovative scientific station to be constructed in the Arctic. Harper proposed a completely new Canadian High Arctic Research Station to be based at Cambridge Bay, Nunavut.

During 2008, as the organization celebrated its fiftieth anniversary, the PCSP was renamed the Polar Continental Shelf *Program*. This change was not envisaged as simply semantic. The base at PCSP Resolute Bay was extended, and joined with new facilities for the Department of National Defence. This construction work was completed in Resolute during 2011. The PCSP remains as a complicated symbol of Canada's scientific sovereignty.

The histories of practice that I have outlined in these pages might allow my readers to be circumspect about this new chapter in the relations between Canadian sovereignty, science, and nationhood. In May 2009, the *Edmonton Journal* reported that PCSP-supported scientists were complaining that, notwithstanding initiatives for the construction of new facilities at Resolute, they had not been allocated enough flying hours to reach their field sites (Munro 2009). As I have argued here, this is the quotidian culture of Arctic field science. Field practice encounters many agents, objects, and obstacles in the high latitudes.

In chapter 1, I argued that post-Confederation Canadian nationalists, desperate for a unitary myth for a physically and culturally diverse polity, envisioned Canada as a northern land where political temperance was to be induced through cultural adaptation to the environment. Such visions of nordicity characterized much pan-Canadian nationalism in the twentieth century, most obviously during the re-election of Prime Minister John Diefenbaker in 1958. Moreover, such imaginative projects were at the same time intertwined with Canadian *geoscientific* traditions. Following the creation of the PCSP, scientists such as Fred Roots attempted to establish an Arctic ethos that stressed the importance of a particular form of field practice: a temperate conduct that focused upon individuals becoming impervious to environmental conditions in order that successful scientific results would be achieved. Not only, as shown in chapter 2, were these ideals difficult to accomplish in scientific practice, they also interacted with debates about the deployment of particular scientific methods in the field. In attempting to introduce field experimental methods, therefore, tensions emerged between how to perform simultaneously as a field scientist and a good Canadian.

In order to examine whether these cultures of temperance were still evident in the activities of the PCSP today, it was necessary to record actual scientific practices. Through ethnography, as discussed in chapters 3, 4, 5, 6, and 7, it became evident that, despite a shift away from direct scientific research to logistical provision, PCSP base managers, and often scientists themselves, continue to try to replicate such cultures by stressing the importance of Arctic scientific practice within discourses of Canadian territorial sovereignty. Arctic science, at least in Canada, remains concerned with being a better Canadian.

However, throughout the episodes examined, Arctic science is also conducted in *intemperate spaces*. As argued in chapters 2, 4, 6, and 7, the accomplishment of experimental manipulation of environments, both historically and in the ethnographic present, is far from trivial. Similarly, ethnographic material showed that social practices of exclusion and emotional reactions to success, failure, and frustration also characterize field science in the Arctic. It is important to note that this does *not* undermine, in any way, the achievements of field scientists and others in the Canadian Arctic, it only makes them all the more remarkable.

It would be easy to argue that the project of detailing hidden voices and social practices is tangential to the study of science in the Arctic, as if science were only constituted through a circumscribed set of activities at a certain number of privileged sites. It is precisely this notion that scientific activity is undertaken only at laboratory-like places that has resulted in many of the intellectual problems of the field sciences. As I have argued, the question revolves around definitions and political commitments. This may have demanded the suspension of everyday, "common-sense" beliefs about what constitutes scientific practice on behalf of some audiences, not least some of the field scientists who acted as my informants, when reading these pages. But I have tried to stress that if practices are understood as arrays of activity performed by *humans*, and centred on shared practical understandings, then it is possible to defer judgement about *which* undertakings are the actually important everyday activities at sites like PCSP Resolute until *after* the prosecution of research. This is always necessary for ethnographic analyses, but it has also been crucial in my historical examinations. In making such judgments, I have been able to present labours and voices whose importance is too frequently disdained. This is a neglect that is re-enforced by an a priori assumption that scientific practices are somehow distinct from the material and emotive contexts of other quotidian activities. It is simply the case that without cooks, mechanics, or pilots, the scientific publications facilitated by PCSP would not exist. As I hope to have shown, the life consequences for those participating in scientific practice, in both historical and contemporary settings, have been stark. Even if it were possible to argue that certain activities are simply logistics, and others are purely science, I have tried to show that it is far too naive to make such distinctions, as ultimately to do so reveals a whole set of prejudices about whose lives and relationships have been of consequence and whose have not. It is not that these are dimensions peculiar to Arctic science, it is more that the logistical challenges make them more obvious in such locations.

To be sure, future investigations of the constitutive processes of Arctic science may find great utility in a Latourian approach that follows a particular research group through the journey from grant application to scientific publication. However, my concern here, as an investigation of the PCSP as an organization created through the quotidian practices of numerous individuals, was a different project. I have been more interested in human exertions rather than those of the inscription or trace. Indeed, the intemperate spaces detailed here are not incidental to scientific practice in the Canadian Arctic; on the contrary, they are absolutely essential to its existence. Without the (often unacknowledged) labours of support staff, competition between researchers, or Inuit participation in research licensing, for example, there would be no fieldwork undertaken by PCSP. It is for this reason that the archival, oral historical, and ethnographic methods utilized in the book have been compatible. This book has argued that ideas of scientific conduct in the field are more complicated when examined through a certain notion of practice. Discourses of scientific nationalism as manifested in the cultures of temperance found in the Canadian Arctic have always been resisted by environmental constituents and human voices and emotions.

It is precisely through investigation of the intemperate spaces of the PCSP that I have had personal anxieties about the proper representation of the Arctic and of those who construct it. For why, exactly, did I become interested in such spaces? Even if it were possible to separate my own personal biography from this question, the ethical quandary of responsibility to my informants would remain. The maintenance of my own reflexive presence and the attempts to give voice to those individuals and issues that could easily have been excluded have therefore been a necessary part of the articulation of the argument.

By focusing upon debates about field observation in Arctic field science, I have attempted to illustrate the complexity of the establishment of the surveyor-as-hero in practice. Moreover, by undertaking my own observations of quotidian activities, I have argued that it is epistemically and ethically impoverished to avoid interpretation in theories of practice. There have been too many recent assertions of the importance of description by social theorists that have underestimated the labours and responsibilities of observation. My point is that this misunderstanding is analogous with that of PCSP field scientists desperate to deploy field experimental methods in the 1960s.

In writing *Studying Arctic Fields*, I wanted to bring to light the conflicting lives that are built through scientific practices in the Arctic. These are often neglected in many commentaries about the future of the region. In some small way, I hope to have shown that understanding global environmental futures will require some grasp of the activities that lie behind those scientific models and experiments. Like all arrays of activity, they involve sets of practices undertaken by humans. Forgetting about those cultural lives, whether in climatic models or in accounts by theorists of scientific practice, will make understanding the future much more difficult. And, ultimately, it will make narratives of the Arctic much less emotionally resonant.

Attempts to know the Arctic have always encountered epistemic anxiety. This book, like the cultural practices outlined within it, is no different. Despite the best efforts of all those associated with the Polar Continental Shelf Program over its nearly sixty-year history, the Arctic continental shelves will always resist cartographic capture and delineation. What has remained throughout, though, is a vision of a possible Canada, developed through scientific field cultures.

NOTE ON METHODOLOGY, SOURCES, AND RESEARCH ETHICS

When the research for this book began, the archives of the PCSP had recently been deposited at Library and Archives Canada (LAC), many of the founding actors were still available for interview, and, crucially, the PCSP continued to operate a High Arctic research station at Resolute, Nunavut. Major archival work was therefore undertaken at LAC, Ottawa, over 2001–05. I was able to request a large amount of information, previously closed to public inspection, under the Access to Information provisions (ATIP) of the Privacy Act, following a formal application (written request made 16 April 2002, permission granted on 11 July 2002).

However, a number of further difficulties presented themselves. The archival files relating to the PCSP were physically transferred to LAC in the early 1990s. However, the original notebook containing the filing codes was lost before (or during) this transfer, meaning that the files held at the LAC are not indexed. The internal LAC information system for archivists further states that the holdings relating to the PCSP had been "dumped in a box" by a member of staff at Natural Resources Canada prior to the transfer. This meant that files were recompiled into random assortments, in the early 1990s, by a student on a two-week work placement designated with the task of categorizing the files.

Moreover, none of the files had been consulted by any member of the public since the introduction of the Privacy Act. Any file requested had therefore to be reviewed by an ATIP officer before it could be released for my consultation. It was impossible for me simply to request all files and work through the material. Instead, I had to cross-reference from secondary literature and oral historical interviews as to where documents pertinent to the research might be contained.

Whilst resident in Canada, and on research visits since, I conducted oral historical research with current and retired PCSP directors, scientists, base managers, mechanics, field technicians, and policy-makers across the country, 2001–05. Locations of interviews included Ottawa, Gatineau, and the surrounding area; Calgary; Iqaluit; Resolute; and Vancouver. Sixty semi-structured interviews were recorded with forty-four individuals. The vast majority of these were tape-recorded, although there were twelve face-to-face interviews and one telephone interview that were not taped. There was some fluidity to these boundaries, with some conversations oscillating between recording and face-to-face, or continuing after the recording had ceased. These interviews ran for between 2 and 4 hours, and occasionally longer, resulting in around 150 hours of recorded tape. The tape-recorded conversations are marked here as (taped). The other interviews that were recorded using hand-written notes are marked as (face-to-face) or as (telephone). These are all listed below.

This research was extended by two field seasons of ethnographic work at the PCSP research base at Resolute, Nunavut, where I observed and recorded the contemporary practices of fieldwork, 2001–02. During this ethnographic component of the research, hundreds of informal conversational interviews were performed and recorded in field notebooks. These conversations are *not* listed individually, for practical reasons of length.

Following the licensing guidelines of the Nunavut Research Institute, I ensured that I had "informed consent" for all participants in this research. I had interviewees sign witness release forms after they were recorded on tape, and discussed confidentiality and possible consequences of participation prior to recording. Throughout the text, as well as below, the identities of those still active in the northern research and policy community have been disguised. As well as pseudonyms, the only information offered is the location of the interview, as it was decided that institutional affiliations would allow the possibility of deciphering the identities of specific individuals. Retired scientists, policy-makers and a serving member of Parliament, as well as a number of other individuals, following the specific wishes of these participants, have been identified by their given names.

Interviews

Adams MP, Peter – House of Commons, Ottawa, 22 May 2002 (taped)
Alexander, Ralph – Resolute, 23 July 2002 (taped)
Alt, Bea – Ottawa, 15 August 2002 (taped; two tapes)
Bacon, Kyle – Ottawa, 29 April 2002 (taped)
Baker, Richard – Resolute, 9 July 2002 (taped; two tapes)
Barmig, Christian – Ottawa, 14 August 2002 (taped; two tapes)
Beckin, Gavin – Ottawa, 19 August 2002 (taped; two tapes)
Brown, Christopher – Resolute, 13 July 2002 (taped)
Cartwright, Aiden – Ottawa, 19 August 2002 (taped)
Collin, Arthur E. – Ottawa, 12 August 2002 (taped)
Cornell, Samuel – Ottawa, 15 April 2002 (taped)
Cutter, Carolyn – Resolute, 21 August 2001 (taped)
Doherty, Nathaniel – Ottawa, 31 March 2001 (face-to-face)
Doherty, Nathaniel – Cambridge, 15 January 2003 (taped)
Errington, Austin – Cambridge, 15 June 2001 (face-to-face)
Errington, Austin – Iqaluit, 24 June 2002 (face-to-face)
Fisher, David – Cornwall (Ontario), 16 August 2002 (taped)
Frisch, Tom – Ottawa, 23 April 2002 (taped)
Hallsworth, Bethan – Resolute, 18 August 2001 (taped)
Hobson, George – Ottawa, 19 April 2002 (taped; two tapes)
Hobson, George – Ottawa, 30 April 2002 (taped; two tapes)
Hobson, George – Ottawa, 2 May 2002 (taped; two tapes)
Hobson, George – Ottawa, 10 May 2002 (taped)
Hobson, George – Ottawa, 15 May 2002 (taped; two tapes)
Howarth, Tristan – Calgary, 4 June 2003 (taped)
Jellings, Sebastian – Resolute, 6 July 2002 (taped; two tapes)
Kent, Heidi – Ottawa, 4 April 2001 (taped)
Kent, Heidi – Ottawa, 22 August 2002 (taped; three tapes)
Koerner, Roy (Fritz) – Ottawa, 9 August 2002 (taped; two tapes)
Koerner, Roy (Fritz) – Cornwall (Ontario), 16 August 2002 (taped)
Lefebvre, Quentin – Resolute, 18 August 2001 (face-to-face)
Loken, Olav – Ottawa, 8 August 2002 (taped)
Lundgaard, Leif – Cornwall (Ontario), 16 August 2002 (taped)
McKinlay, Martin – Resolute, 21 August 2001 (taped)
Merton, Alan – Resolute, 13 July 2002 (taped; two tapes)
Mitchell, Isabel – Cambridge, 18 June 2001 (face-to-face)

Nassichuk, Walter – Calgary, 4 June 2003 (face-to-face)
Phillips, Caroline – Cornwall (Ontario), 16 August 2002 (taped)
Plummer, Michael – Resolute, 9 July 2002 (taped)
Postan, Catherine – Resolute, 29 July 2002 (taped)
Presley, Bill – Ottawa, 13 August 2002 (taped)
Rigby, Bruce – 10 May 2001 (telephone)
Rigby, Bruce – Iqaluit, 25 June 2002 (taped)
Roots, E.F. (Fred) – Gatineau, 22 April 2002 (taped; two tapes)
Roots, E.F. (Fred) – Gatineau, 6 May 2002 (taped)
Roots, E.F. (Fred) – Gatineau, 13 May 2002 (taped; two tapes)
Roots, E.F. (Fred) – Gatineau, 17 May 2002 (taped; two tapes)
Rowley, Graham – Ottawa, 18 April 2002 (face-to-face)
Scarlett, Jared – Resolute, 21 August 2001 (taped)
Smith, David – Vancouver, 10 November 2000 (face-to-face)
Smith, David – Vancouver, 7 September 2001 (taped)
St Onge, Denis – Ottawa, 11 April 2002 (taped; two tapes)
Stager, John – Vancouver, 14 March 2000 (face-to-face)
Stager, John – Vancouver, 7 November 2000 (face-to-face)
Stager, John – Vancouver, 12 April 2001 (face-to-face)
Stager, John – Vancouver, 6 September 2001 (taped)
Suedfeld, Peter – Vancouver, 10 April 2001 (face-to-face)
Thomas, Mary Ellen – Iqaluit, 25 June 2002 (taped)
Thornsteinsson, Ray – Calgary, 4 June 2003 (taped)
Warner, Cheryl – Resolute, 27 July 2002 (taped)

Primary Sources

National Archives of Canada, Ottawa
Record Group 22
    Volumes 544–5
Record Group 45
    Volumes 300–6, 310–14, 316–25, 328, 330–3, 335–7, 339–42, 344–5, 351
National Library of Canada, Ottawa
House of Commons Debates. Official Reports

NOTES

INTRODUCTION

1 The manuscript draws from over thirty-six months of ethnographic fieldwork, involving archival consultation, oral histories, semi-structured interviews, and participant observation, conducted over the last decade or so at sites across Canada and elsewhere.

2 The PCSP was renamed the Polar Continental Shelf Program in 2008, marking its fiftieth anniversary. Throughout out the text, I use *Project* to refer to the organization in historical or ethnographic context, and use *Program* to refer to the organization today.

3 The PCSP has already attracted its own historians. One of the former directors has produced his own brief, personal reflection (Hobson 1990). A photographic history and VHS video were produced in 1986 for the twenty-fifth anniversary of the organization (Foster and Marino 1986; Energy, Mines and Resources Canada 1986). Inevitably, however, these in-house publications take little account of the wider scientific and political context.

4 For reference, the intended scientific field season for projects operating out of PCSP Resolute in 2002 was 23 March to 10 September, with days shaved off either end as projects were cancelled. Each year the operational season tends to get shorter as the result of financial constraints on the PCSP budget. This gradual decrease in the length of the season, and thus in the annual employment for employees on seasonal contracts, has resulted in serious financial hardships for particular families.

5 The few who are not offered PCSP support for whatever reason tend to stay in one of the two hotels in the hamlet of Resolute, the Co-op Hotel (Inns North) or the South Camp Inn, or at the Narwhal Hotel near the

airport. The Narwhal Hotel is normally used by companies with personnel employed on business in or around the airport complex, whereas the hamlet hotels are usually used by scientists, tourists, and other visitors.

6 Some researchers, however, use PCSP Resolute for daily accommodation, making day trips to field sites on Cornwallis Island.

7 Interview with Peter Adams MP, House of Commons, Ottawa, 22 May 2002.

CHAPTER ONE

1 The classical argument, by Northrop Frye (1971), is that Canadian literature has been broadly concerned with defining *place*, whereas American literature has been concerned with discovering individual identity.

2 In July 1956, Nasser nationalized the Suez Canal, leading to the invasion of Egypt by Israel, France, and the United Kingdom on 29–30 October. A withdrawal of the invading forces was negotiated by the United Nations, with deployment of UN peacekeeping forces in November 1956 (Whitaker and Marcuse 1994).

3 The Progressive Conservatives won 112 seats to the Liberals' 105, with the Co-operative Commonwealth Federation (CCF) taking 25 seats, and the final 19 seats won by Social Credit.

4 This was the first majority for the Conservatives in Quebec since 1887. Diefenbaker had struck a successful electoral alliance with the Union Nationale, a French-Canadian nationalist party.

5 The reasons for any electoral success are, of course, multiple and complex. When Lester Pearson became Liberal leader at the party convention in January 1958, his manner of acceptance was deemed by some present to be overly arrogant and presumptive. Morton argues that this contributed to the unpopularity of the opposition, and that it was as much this, rather than any populist appeal, that led to the Diefenbaker landslide (Morton 1997).

6 The use of Tennyson's "Ulysses" appears as a direct echo of British imperialistic attempts to harness polar activity for national goals. "To Strive, To Seek, To Find, And Not To Yield" was also the inscription used, at the suggestion of expedition member Apsley Cherry-Garrard, on the memorial cross for the deceased members of the Scott expedition to the South Pole, erected at the summit of Observation Hill in January 1913 (Jones 2003).

7 The national dream associated with the Northern Vision was quickly extinguished, however, because of the short saga of the Avro Arrow, "a

great Canadian tragedy" (Whitaker and Marcuse 1994, 155). In the early 1950s, the Cabinet Defence Committee approved a $27 million contract with Avro Arrow Limited to develop two prototypes for a new fighter-interceptor to defend against bomber aircraft attacking from over the North Pole. However, the number of aircraft required for the RCAF was overestimated, and mooted sales to the United States and United Kingdom were unrealistic, both for political reasons related to the impact on the domestic defence industries, and because the Arrow was "designed around Canada's peculiar geographic needs," having features such as an expensive second engine to guarantee against engine failure over the Canadian north (Isinger and Story 1998, 47). Moreover, Orenda Engines Limited also began to develop a PS-13 Iroquois engine, because the RCAF were not satisfied with existing foreign engines. Massive cost overruns meant that the Diefenbaker government was forced to confirm what the outgoing Liberals had privately intended. Financial issues notwithstanding, the launch of *Sputnik* completely changed the nature of Cold War defence strategy, as fighter-interceptors became immediately less important. The stark decision was taken on 20 February 1959 to cancel the Arrow and Iroquois. As Avro and Orenda were both divisions of A.V. Roe Canada Ltd, over 14,000 workers were laid off at the plants in Malton, Ontario, and many highly skilled engineers left Canada for the United States and United Kingdom. The six existing Arrows and thirty-one others in various stages of completion were stripped of classified material and blow-torched into scrap metal, as was standard operating procedure at the Department of Defence. What is important about this Arrow contestation for our purposes is not so much the details, but the fact that the new sense of nationhood that Diefenbaker had exploited for election was dead within twelve months.

8 Relatively few studies have discussed the results of the Northern Vision. The work that does exist has tended to investigate the repercussions of this policy vision for Inuit, Inuvialuit, and other indigenous groups of Northern Canada (Abele 1987; Ambrose 2005).

9 This is evident in the House of Commons speech by the leader of the opposition, Lester Pearson, on 14 August 1958 (House of Commons 1958).

10 Alvin Hamilton, minister of northern affairs and national resources, memorandum to Cabinet, "Canadian Activities in the Polar Basin," 3 April 1958, Library and Archives Canada (LAC), Record Group (RG) 45, vol. 300, Cabinet document 58/58 [marked "Confidential"].

11 Ibid., 1.

12 Ibid.

13 The DEW Line finally became operational on 31 July 1957, just before the threat of ICBMs became apparent to North American military strategists later that fall.

14 Interview with George Hobson, Ottawa, 19 April 2002. Although this transition by the US Navy's nuclear-powered submarine *Nautilus* was well reported at the time, it has only been over the past decade or so that histories of submarine activity during the Cold War have been published (Reed 1958; Sontag and Drew 1998).

15 Interview with Arthur E. Collin, Ottawa, 12 August 2002.

16 Hamilton, "Canadian Activities in the Polar Basin," 1.

17 Ibid.

18 See, for example, G.W. Rowley, "Canadian Sovereignty and Arctic Waters," 1959, LAC, RG22, vol. 545, file Rowley – Advisory Committee on Northern Development 1959, ACND Material, Jan 1, 1959 – April, 30, 1959. There had also been anxiety over these issues under the previous Liberal government; G.W. Rowley, "Sovereignty in the Canadian Arctic," 1954, LAC, RG22, vol. 544, file Rowley – Advisory Committee on Northern Development 1954, Part 3 – December 18, 1954 – January 29, 1953 [*sic*].

19 Prior Canadian activity in the region of the shelf was limited to Stefansson's Canadian Arctic Expedition, 1913–18, and the Geological Survey of Canada's recent Operation Franklin, 1955.

20 However, UNCLOS I failed to achieve universal ratification, leading to further conferences in 1960, 1973, and 1982 (Steinberg 2001).

21 G.W. Rowley, "Scientific Activities in the Canadian Arctic," 1953, LAC, RG22, vol. 544, file Rowley – Advisory Committee on Northern Development 1953, Part 1 – July 23, 1953 – December 31, 1953.

22 "A Research Program to Provide Information on the Extent and Characteristics of the Continental Shelf and Other Phenomena, Both Physical and Biological, of the Polar Basin – Polar Continental Shelf Project," report of Technical Sub-Committee of the Advisory Committee on Northern Development, 10 March 1958, LAC, RG45, vol. 300, file 1-1-1 [marked "Confidential"].

23 W.E.O. Halliday, registrar of Cabinet, Privy Council Office, record of Cabinet decision, meeting of 22 May 1958 – "Canadian Activities in the Polar Basin," 29 May 1958, LAC, RG45, vol. 300, Cabinet document 126-58 [marked "Confidential"]; handwritten memorandum by Don [Cameron, PCSP administrative officer] to G. Hobson, director of PCSP, "Attempt to Trace Changes in Geographical Ambit of the Polar Continental Shelf Project in the Period 1959–73," (circa 1973), LAC, RG45, vol. 304, file 1-8-2.

24 During Cabinet discussions, the PCSP was transferred to control of the ministry of Paul Comtois, the Department of Mines and Technical Surveys; Paul Comtois, minister of mines and technical surveys,

memorandum to Cabinet: "Canadian Activities in the Polar Basin," 15 May 1958, LAC, RG45, vol. 300, Cabinet document 126-58 [marked "Confidential"]. This dispute was over the fact that the original department of Alvin Hamilton, Northern Affairs and National Resources, lacked any scientific expertise within its staff; interviews with George Hobson, Ottawa, 19 April, 30 April, 2 May, 10 May, 15 May 2002.

25 E.F. Roots, "Notes on the Organization, Objectives, and Operation of the Polar Continental Shelf Project, Department of Mines and Technical Surveys," 10 August 1961 (statement prepared for Royal Commission on Government Organization), LAC, RG45, vol. 304, file 1-7-1.

26 Comtois, "Canadian Activities in the Polar Basin."

27 Roots, "Notes on the Organization, Objectives, and Operation of the Polar Continental Shelf Project," 2.

28 W.E. van Steenburgh to R.G. Robertson, chairman, Advisory Committee on Northern Development, Department of Northern Affairs and National Resources, 12 December 1958, LAC, RG45, vol. 304, file 1-7-8 [marked "Confidential"], 1.

29 Ibid., 2, 4.

30 E.F. Roots, co-ordinator, PCSP, "Comments on Mr Gray's reply to Mr Robertson's memo," 16 December 1958, LAC, RG45, vol. 304, file 1-7-8, 2–3, 4.

31 Interviews with E.F. Roots, Gatineau, 22 April, 6 May, 13 May, 17 May 2002.

32 Interview with Denis St-Onge, Ottawa, 11 April 2002.

33 Interview with Collin, 12 August 2002. This notion of a golden age of fieldwork was evident in interviews with many other retired Arctic scientists. As geologist Tom Frisch puts it, "I'm really out of that now, with being retired. It's not something I'm really involved with right now. I'm just lucky I had an opportunity to work in the Arctic. In the good old days." Interview with Tom Frisch, Ottawa, 23 April 2002.

34 Interview with St-Onge, 11 April 2002.

35 The influential sociologist Robert Merton argued that four "norms" comprised the ethos of scientific practice: universalism, communism, disinterestedness, and organized skepticism (Merton [1942] 1973).

36 Interview with Hobson, 19 April 2002.

CHAPTER TWO

1 Interview with St-Onge, 11 April 2002.

2 These debates amongst physical geographers drawn from here are particularly apposite disciplinary instantiations of much wider discussions across the philosophy of earth and environmental science that first

surfaced in the early 1960s, as the multidisciplinary PCSP field scientists began their research (Simpson 1963; Watson 1966).

3 Important historical and anthropological work on wildlife management in northern Canada has stressed the politics around the co-production of knowledge between scientists and indigenous peoples (Sandlos 2007; Nadasdy 2008). Philosopher Jeff Kochan (2015) has recently attempted to classify these styles of field science in the Yukon and Northwest Territories.

4 The biographical details here are drawn from a series of interviews with Roots, 22 April, 6 May, 13 May, and 17 May 2002. A biographical sketch of Roots is also offered by Beeby (1994) in a discussion of his work in Antarctica.

5 On one occasion, Roots had to act as assistant in a surgical operation to remove an infected eye from a fellow geologist, using iron scraps and a welding rod (Beeby 1994).

6 Roots was later awarded the Patron's Medal of the Royal Geographical Society, London, in June 1965. *Gazette*, 6 April 1965, LAC, RG45, vol. 337, file 2-1-7.

7 "Research Program to Provide Information on the Extent and Characteristics of the Continental Shelf," 10 March 1958.

8 A hallmark of the organization's success was that, following completion of the study of the continental shelf area, the geographical extent of PCSP research grew to encompass the entire Canadian Arctic by the late 1970s.

9 E.F. Roots, co-ordinator PCSP, to W.E. van Steenburgh, director-general scientific services, Department of Mines and Technical Surveys, 7 August 1959, LAC, RG45, vol. 300, file 1-1-3.

10 Handwritten notes by E.F. Roots for PCSP Steering Committee Meeting, 16 February 1960, LAC, RG45, vol. 300, file 1-1-3, 4.

11 E.F. Roots, "Notes on Ice Studies Conducted by the Polar Continental Shelf Project, Department of Mines and Technical Surveys," paper no. 32, 4 October 1960, LAC, RG45, vol. 308, file 3-8-0, 1.

12 Copy of letter from K. Arnold, Meighen Island Icecap, to E.F. Roots, 11 July 1959, LAC, RG45, vol. 308, file 3-8-0, 1.

13 Interview with St-Onge, 11 April 2002.

14 Ibid.

15 Interviews with Christian Barmig, Ottawa, 14 August 2002; Leif Lundgaard, Cornwall, 16 August 2002; Bill Presley, Ottawa, 13 August 2002.

16 E.F. Roots, co-ordinator, PCSP, to R.B. Code, chief, Personnel Division, Department of Mines and Technical Surveys, 31 August 1959, "[Name withheld], Labourer, Acting as General Assistant to Polar Continental Shelf Project," LAC, RG45, vol. 302, file 1-2-1, 1.

17 Ibid.

18 Ibid.

19 "Canada Will Do Research in the Polar Basin," Department of Mines and Technical Surveys, Government of Canada, 20 February 1959, LAC, RG45, vol. 306, file 2-9-1, 1. Note the overt masculinist discourses circulating through this statement.

20 This claim about PCSP fieldworkers being the first human presence on Ellef Ringnes Island and Meighen Island was also made in a number of interviews with protagonists (interviews with St-Onge, 11 April 2002; Hobson, 15 May 2002).

21 Interview with Roots, 6 May 2002.

22 E.F. Roots to Computing Devices of Canada Ltd, "Attention: Mr Peter Wilson," 1 June 1967, LAC, RG45, vol. 374, file 4750, pt 1. The Decca 6f Lambda chain was the first Decca system that had *survey*, as opposed to aircraft-navigation, capability (interview with Roots, 22 April 2002).

23 [Bert Burry] to George [Burry] and Jimmy [Burry], 12 April 1967, LAC, RG45, vol. 377, file 5155-4, pt 1, 1.

24 Ibid., 4.

CHAPTER THREE

1 It turned out that the PCSP base was not visible from the terminal, as other buildings obstructed the line of sight.

2 Interview with Bea Alt, Ottawa, 15 August 2002.

3 All-terrain vehicles (ATVs) are four-wheeled machines that can carry a researcher across Arctic terrain to field sites.

4 Important correspondence and reports are often "sent south" to Ottawa to be dealt with by the PCSP director.

5 The standard rate (2002) for hotels in Iqaluit was around $200/night.

6 Charles used to be employed as a base janitor and had been responsible for base upkeep in previous seasons, initially as an unpaid volunteer. He then worked in the Resolute office, before working for four seasons at PCSP Eureka.

7 Although this is a colloquial term and has much more complicated origins than simply a form of racist abuse, I use "Newfoundlander," except when recording conversations with others, in order to guard against accusations of insensitivity. One reason that the seasonal, technical staff are all from Newfoundland is related to the collapse of the regional economy of Atlantic Canada, following the introduction of quotas in the cod fishery in the 1990s. Economic hardships in Newfoundland mean that these individuals are willing to take summer seasonal work in the Arctic as

their only annual employment. However, some of the Newfoundlanders at PCSP have spent their entire working lives in the north, and the relationship between migratory labour from this province and Arctic science and industry has deeper roots.

CHAPTER FOUR

1 Conversation with Edward Freeman, Resolute, 20 August 2001, recorded in field notes, August 2001, book 1.
2 Conversation with Heidi Kent, Resolute, 31 July 2002, recorded in field notes, June–August 2002, book 3.
3 Ibid.
4 Interview with Quentin Lefebvre, Resolute, 18 August 2001, recorded in field notes, August 2001, book 1.
5 Ibid.
6 Ibid.
7 Incidentally, connection speeds for the Internet were much faster in 2002 than in 2001.
8 This would slow the download speed of satellite photographs and weather reports in the base office downstairs.
9 This may be an incorrect perception, a simple artifact of my self-consciousness as I returned to PCSP Resolute for a second period of fieldwork.
10 Conversation with Andrew Walder, Resolute, 3 July 2002, recorded in field notes, June–August 2002, book 1.
11 Ibid. Russell Merris is a septuagenarian British botanist who wanders unsupported around the High Arctic collecting plants, especially on Ellesmere Island. He catches flights with spare room, informally, into and out from the field. He is regarded as both a local legend and an amusing eccentric, especially by the old-timers in Resolute, because he has made this annual trip for years, whilst supposedly dressed in a lounge suit and tie, and has always returned unscathed. However, in August 2001 Merris missed the flight back to Resolute that he had planned to intercept, and had to trigger an emergency beacon, causing serious logistical problems for the RCMP and PCSP in Resolute.
12 Conversation with Walder, 3 July 2002.
13 Ibid.
14 Conversation with Edward Freeman, Resolute, 24 August 2001, recorded in field notes, August 2001, book 2.
15 Ibid.
16 It appeared that although Bradley's had failed to secure the PCSP contract in 2001, they had found alternative High Arctic work through NASA.

17 This model of polar field practice revolves around the veneration of "good ice men" who could perform tasks effectively under hostile conditions and extreme cold. Research by Klaus Dodds (2002) indicates that these traditions were also prevalent in the Falklands Islands Dependencies Survey, a predecessor of BAS, during the 1950s.
18 Interview with Richard Baker, Resolute, 9 July 2002.
19 The Narwhal Hotel is operated by Narwhal Arctic Services, a division of Frontec. This company also runs the annual sealift for the various organizations in Resolute.
20 Conversation with various scientists at lunch, 3 July 2002, recorded in field notes, June–August 2002, book 1.
21 Ibid.
22 Indeed, I have found that it is *my* particular polar bear story that most interests scientists and non-scientists alike about my fieldwork.
23 I approached these two women at my first meal at Resolute for no conscious reason and yet, subconsciously, I likely approached them first as they sat alone and perhaps appeared less threatening. Within the status hierarchy at PCSP Resolute, these individuals occupied a relatively lowly position as botanists employed by a government department. In turn, however, this chance encounter led to many future introductions in Resolute and later when conducting oral historical interviews in Ottawa. There is a large ethnographic literature documenting similar cases to this instance of the phenomenon of the new outsider.
24 It is therefore interesting that both field sites were not only close to Resolute, both were close to "dirt" roads and at sites that already have had significant human presence: a patch of grass just behind the hamlet and an archaeological site that has been restored for visitors.
25 Some of the workspace in this lab was taken up with the PCSP's collection of large-scale maps of the Arctic islands.
26 Interview with Lefebvre, 18 August 2001.
27 An alternative spirit level had to be flown up from Resolute later in the week.
28 I later visited Leif in Cornwall, outside Ottawa, and was fascinated by his tales of science in the field.
29 Indeed, it is not necessary to be fully "able-bodied" to be a successful field scientist in the Canadian Arctic.

CHAPTER FIVE

1 *Qallunaat*, translated literally, means the "big, high (eye)brows" in Inuktitut, and is used to refer to non-Inuit (Laugrand and Oosten 2002).

2 Iqaluit is the territorial capital of Nunavut and was known as Frobisher Bay until 1987.

3 For full accounts of the relocations to Resolute, see the discussions in Tester and Kulchyski (1994) and Marcus (1995).

4 Interview with Ralph Alexander, Hamlet Council offices, Resolute, 23 July 2002.

5 Interview with Graham Rowley, Ottawa, 18 April 2002.

6 Interview with Alexander, 23 July 2002.

7 Although the vast majority of research groups operating out of PCSP Resolute have adapted to this situation, some field teams still do not fully conform to these new licensing arrangements. One postgraduate student, for example, told me that they had been unable to fly to their intended licensed site, so the party had returned to a camp that the lead professor had used during previous field seasons.

8 Interview with Alexander, 23 July 2002.

9 Ibid.

10 Conversation with Edward Freeman, Resolute, 26 July 2002, recorded in field notes, June–August 2002, book 2.

11 Interview with Alexander, 23 July 2002.

12 The generally relayed argument is that Inuit catch up on sleep during the winter.

13 My positionality in this cultural space was obviously complicated. Among natural scientists at Resolute, I was positioned as an outsider by being a social scientist. Among Inuit, *all* researchers are held to be outsiders. This situation was further complicated by my English identity – my country of origin has been historically connected with colonial practices in the Canadian Arctic. However, at the time of conducting fieldwork, I had recently graduated, with a master's degree from a leading western Canadian university and had been resident in Ottawa for a further eighteen months. This moderated my status somewhat.

14 Interview with Bruce Rigby, Nunavut Research Institute, Iqaluit, 25 June 2002.

15 Interview with Mary Ellen Thomas, Nunavut Research Institute, Iqaluit, 25 June 2002.

16 Interview with Rigby, 25 June 2002.

CHAPTER SIX

1 Beatty (2005a) provides an effective review of recent anthropological writing on emotion and affect. For a comprehensive survey of anthropological

work that investigates the range of devices that convey affective meaning, see Besnier (1990).

2 There is, of course, a literature on the ethical challenges of emotional fieldwork in geography (Widdowfield 2000; Powell 2002; Bondi 2003). Much of this research has, however, concentrated on the emotional consequences of conducting qualitative interviews rather than participant observation. Drawing from object relations psychoanalysis, for example, Bondi (2003) discusses the utility of the concept of *empathy* for qualitative interviews. Although this provides for stimulating opportunities to rethink the research relationship, I would argue that ethnographic access is negotiated at the community level and thus can provide for slightly different encounters.

3 Briggs (1987) has noted that any connections between her field research and psychoanalytic theory emerged *retrospectively* – ethnographic observation of social interactions resulted in her interpretations, rather than prior reading. It is important to note, though, that Briggs's person-centred ethnographies, developed through field experiences, appear to have important epistemic parallels with those psychoanalytical and psychotherapeutic approaches to emotional geography developed by Bondi (2003).

4 The seasonal morphology of Inuit society was first outlined by Marcel Mauss (1906).

5 Christmas is translated as *quviasuvik* (literally "time of happiness") in Inuktitut (Stuckenberger 2005).

6 Conversation with Yvonne Norton, Resolute, 11 August 2001, recorded in field notes, August 2001, book 1.

7 Conversation with Edward Freeman, Resolute, 19 August 2001, recorded in field notes, August 2001, book 1.

8 Conversation with Barry Fuller, Resolute, 26 July 2002, recorded in field notes, June–August 2002, book 2.

9 This was due to the initial conditions of my presence at the base, in that I had myself applied successfully for PCSP support as a scientific researcher. Although conducting ethnography of science, as an individual with PCSP support, I was sometimes constructed as a "scientist" like any other researcher on base.

10 Riches (1977) identified similar tendencies for the development of social cliques and rituals, involving festivities and drinking, amongst Euro-Canadians in settlements in northern Canada during the 1970s.

11 Conversation with Freeman, 24 August 2001.

12 Ibid.

13 Conversation with Trevor Bennett, Resolute, 21 August 2001, recorded in field notes, August 2001, book 1.
14 Interview with Lefebvre, 18 August 2001.
15 Conversation with Freeman, 24 August 2001.
16 These competing hotels are viewed warily by the PCSP management. As Edward argues, "Foreign scientists could be supported by Polar Shelf, but we must be careful not to take business from the village. We have to fill the hotels in Resolute first." However, only one hotel is actually run by Inuit.
17 The PCSP Ice Island was in operation circa 1986–89, and was literally a floating base on the Arctic Ocean from which scientific projects were undertaken. The Newfoundlanders made constant reference to those individuals who had been employed on the Ice Island, and for how long they had "served." This was partly because it was seen as the ultimate challenge in Canadian Arctic employment, but it could also be interpreted in a sometimes humorous way, as it was generally regarded amongst PCSP staff and scientists that the Ice Island episode was a huge failure. Although a twenty-five-year program of research had been envisaged, the chosen island disintegrated within three seasons.
18 Conversation with Andrew Walder, Resolute, 25 August 2001, recorded in field notes, August 2001, book 2.
19 This policy of sending staff home early for financial reasons has always been a management strategy for PCSP. The first PCSP co-ordinator, E.F. Roots, for example, sent a telegram to Lionel Laurin, the camp manager at Tuktoyaktuk, in May 1969: "If we are to start operations by Feb. 1 of next year it is imperative that you reduce your total base camp staff by six men months before August 15. This would mean [a] reduction of two men for three months or equivalent. We are also reducing casual employment in other areas." E.F. Roots to L. Laurin, PCSP Tuktoyaktuk, telegram, 1.30 p.m., 22 May 1969, LAC, RG45, vol. 322, file 5-1-1, pt 1, Radio communications – Miscellaneous, Part 1, 1968–1969.
20 See Doing (2004) for similar findings regarding the politics of labour in an ethnographic study of a physics laboratory.
21 Conversation with Barry Fuller, Resolute, 26 July 2002, recorded in field notes, June–August 2002, book 2.
22 Conversation with Barry Fuller, Resolute, 21 August 2001, recorded in field notes, August 2001, book 1.
23 Ibid.
24 Conversation with Fuller, 26 July 2002.
25 Ibid.

26 Kristin worked at Tuk for eighteen years before "being dragged over" to Resolute. A number of base staff had worked together at PCSP Tuk before it was shut down, and I think a lot of the resentment directed towards some individuals in senior management results from this strategic decision. Kristin had heard rumours that Tuk might be reopened for the 2003 season and was very excited about this possibility.

27 There have been few social studies that investigate the ethical implications of representing such spaces. However, in organizational studies, Michael Rosen has undertaken important work (Rosen 1985, 1988). Through an analysis of the social drama of the Christmas party organized for all staff at Shoenman and Associates, Rosen argues that alcohol holds an important social function. Enacting the "play" of the party, for Rosen, suggests the importance of those moments for peculiar agency beyond normal controls that, at the same time, reproduce the very social structures of the organizational community (Rosen 1988).

CHAPTER SEVEN

1 Malcolm Ramsay interviewed in *Vets in the Wild: Polar Bear Special* (BBC 2000). This was a televised interview by BBC journalists with Professor Ramsay. As I will discuss, this was recorded a couple of days before he died tragically in a helicopter crash near Resolute.

2 There were a number of social connections between certain graduate students, pilots, and some base staff. This meant that some graduate natural scientists saw themselves as "local experts" on PCSP, having been told of various tales and rumours. This knowledge was sometimes used to assert a form of informal social superiority over natural scientists who did not associate socially with pilots and base staff.

3 As they were journalists, however, the BBC filmmakers had to stay in one of the hamlet hotels.

4 In retrospect, I sometimes wonder whether, during my first couple of days at Resolute and before I had had a chance to talk to everyone, some scientists thought I was about to write an "exposé" of PCSP.

5 Conversation with Andrew Walder, Resolute, 14 July 2002, recorded in field notes, June–August 2002, book 2.

6 Conversation with Andrew Walder, Resolute, 6 July 2002, recorded in field notes, June–August 2002, book 1.

7 Ibid.

8 Conversation with Walder, 14 July 2002. PCSP has supported documentary films such as *An Arctic Puzzle* (Department of Fisheries and Oceans

1993); *Mars on Earth: Preparing for Life on the Red Planet* (LaRose 2000); *Search for a "Tropical Arctic"* (Levy 1988); and *Secrets of Ice* (Low 1991).

9 Another example of the problems of cross-departmental communication was during a conversation between a base manager and a visiting Canadian Coastguard officer. NRCan, the host department of PCSP, had bought some Iridium satellite telephones for trial, and the Canadian Coastguard did not know about this, despite planning their own trials. As Edward told the Coastguard, the satellite telephones "are the greatest thing since GPS up here."

10 The Canadian Rangers are a branch of the Canadian Armed Forces, mainly composed of indigenous peoples from the northern regions of Canada, such as Inuit and Inuvialuit. They help provide a northern military presence, undertaking basic surveillance duties in the vast northern territories whilst hunting. This role was, obviously, much more important during the Cold War period. Some scientists complained that this Ranger expedition had ruined the PCSP Skidoos, used for snow-melt surveys in the spring, and that the parts from three machines had been needed to get a single Skidoo operational.

11 There was a miasma around the base that the PCSP years were drawing to a close. Edward noted at lunch one day, "A decade ago we had three Twin Otters and twelve helicopters, and it was 'go, go, go.'"

12 There was a PCSP base manager at Eureka for part of the 2001 season, but, likely for financial reasons, not in 2002.

13 Conversation with Freeman, 24 August 2001. Edward believes that Charles will not stay with PCSP, given his qualifications, although he could be a future director, and that Yvonne could also leave.

14 Bradley's had previously won the PCSP contract and their pilots thus have significant High Arctic experience, but had not operated in Antarctica. Bradley's is the name "old-timers," such as Edward, still use for First Air.

15 Conversation with Freeman, 24 August 2001. Interestingly, Edward's list of government departments neglects to mention Natural Resources Canada and Transport Canada.

16 My impression is that this is a Catch-22 situation. Scientists are relocating research away from the Western Arctic because they do not think there is any potential for PCSP support, which means that PCSP applications will be down, so there will not be any evident demand. This is the problem of a logistics organization that supports projects decided on separately by research councils or other government departments.

17 Conversation with Freeman, 26 July 2002.

18 Conversation with Freeman, 24 August 2001.

19 That is, after 11 September 2001 and the changing geopolitical situation.
20 Conversation with Freeman, 26 July 2002.
21 Conversation with Freeman, 24 August 2001.
22 Ibid.
23 Interestingly, despite a lack of public knowledge of PCSP, there would be various distinguished visitors to PCSP over a season. In July 2002, David Anderson (minister for Environment Canada) headed a group that visited the base, including Michael Meacher (then minister for Environment, UK), the Norwegian minister for Environment, the Austrian ambassador to Canada, and the British high commissioner to Canada.
24 Conversation with Freeman, 26 July 2002.
25 Ibid. The GSC recently moved an office to Iqaluit, creating the Canada-Nunavut Geoscience office, and the relocated federal employees applied for PCSP support for fieldwork on central Baffin Island. However, although PCSP provided over half of the flying hours needed for the operations of the Canada-Nunavut Geoscience Office in 2002, it was apparent that members across departments of the Government of Nunavut in Iqaluit were unaware of this in interviews and conversations that I conducted in June 2002. Various branches of the Government of Nunavut thus appeared to be sometimes unaware of the important federal subsidy provided by PCSP for field scientific activities.
26 Conversation with Freeman, 24 August 2001.
27 These could, of course, be a consequence of the junior status of such staff, rather than gender. However, junior male members of staff were not teased in this way.
28 Interestingly, the scientist found out, after the rest of her group had left, that the researchers already at this field site had not answered on sked that morning because they were chasing away a polar bear. She believed that she had therefore made the correct decision.
29 When helping this lecturer in the field, on different occasions, I was also struck by the numerous times that she remarked that it was cold and that, when we were finished, she was very happy to be able to return to a warm building rather than a tent.
30 Conversation with Alan Merton, Resolute, 13 July 2002, June–August 2002, book 2.
31 This same scientist, for example, had applied for research funding from the Nunavut Department of Heritage, Cultures, Elders and Youth, but after applying to Iqaluit, the form needed a signature from Igloolik. The application was sent in both directions by expensive courier, but "if 'Joe Clerk' had signing authority," he argued, there could have been significant reductions in overall costs.

32 When I asked this scientist about changes in future scientific agendas and desires for more holistic research in the Arctic, she replied, "Have you been talking to Bruce Rigby?" Rigby was then the director of the Nunavut Research Institute in Iqaluit.
33 I observed another researcher asking Edward if there was an official document outlining how flying hours are allocated and divided. This was the researcher's first season through Resolute, and he "got shirty," as Edward put it, over the radio when he found out that half his flying hours were used up. Edward stated that flying hours are divided up on the basis of payload, flying time, and other considerations. This dispute apparently led to a statement concerning the allocation of flying hours being published on the PCSP website in autumn 2002.
34 As noted before, this campaign was ultimately unsuccessful, with the name of the organization changed to Polar Continental Shelf Program in 2008.
35 I do not believe that this procedure is quite as cynical as it might appear, but Inuit participation is certainly now used as a "competitive advantage" in attempts to secure funding from limited pots.
36 In this context, "ground-truthing" refers to first-hand observation by a field researcher to elucidate patterns previously revealed by remote sensing methods.
37 Julian, the researcher from Water Survey Canada, believed that "lots of people began their careers up north because they were trying to escape from the south or just trying to make money. They should never have come up. Drink and drugs were used to escape the monotony at places like Alert, Eureka, and the Barren Lands. Most of those same people are now fried, because they can't readjust socially to the south."
38 At this point, I wondered whether these scientists might have been "world class in Canada" in the same way that Mordecai Richler taunts Canadian novelists for being "world famous in Canada." This is a well-known jest in Canadian letters (Atwood 2003).
39 Although in many ways both practices continued to coexist. Certainly, scientist Cheryl Warner advises all her students to write down everything they observe in the field, as well as all the ideas they have about what is happening at the site, in their field notebooks. The conception of the "eye" in the field is associated most with the British geomorphologist S.W. Wooldridge.
40 Moreover, Canada became a global leader in permafrost research in the 1950s and 1960s as Mackay undertook field experiments at exactly the time that PCSP was established. As the philosophy of field science has

changed, to incorporate indigenous knowledge, it is increasingly difficult for Canada to maintain its position.

41 This is significant, as the men's bathroom in the beaker building occasionally acts as a "space of dissent" as, like the field sites, it is less obviously under the surveillance of the base staff.

42 The most excluded, of course, are those unsuccessful in applications for PCSP support. Some scientists noted that many northern journalists are now excluded from PCSP Resolute.

43 There was a similar attitude amongst most people on base towards French Canadians. During sked conversations with research groups from Quebec, Andrew would sign off with a "Mercy" in a purposefully anglicized way. Although Yvonne initially trained as a linguist, is fluently bilingual in Canada's official languages, and often sings in French and English interchangeably, and Charles is also very proficient in French, the majority of the PCSP base staff, such as the Newfoundlanders, are avowedly anti-French, and most scientists are not much more accommodating. I do not believe this is conscious behaviour, just a result of the constitution of the community of scientific practice.

REFERENCES

Abele, F. 1987. "Canadian Contradictions: Forty Years of Northern Political Development." *Arctic* 40:310–20.

Ahnert, F. 1980. "A Note on Measurements and Experiments in Geomorphology." *Zeitschrift für Geomorphologie Supplementband* 35:1–10.

Althoff, W.F. 2007. *Drift Station: Arctic Outposts of Superpower Science*. Washington, DC: Potomac Books.

Amagoalik, J. 2000. "Wasteland of Nobodies." In *Nunavut: Inuit Regain Control of Their Lands and Their Lives*, edited by J. Dahl, J. Hicks, and P. Jull, 138–9. Copenhagen: International Working Group for Indigenous Affairs.

Ambrose, L.M. 2005. "Our Last Frontier: Imperialism and Northern Canadian Rural Women's Organizations." *Canadian Historical Review* 86:257–84.

Anderson, B. 1991. *Imagined Communities: Reflections on the Origin and Spread of Nationalism*. Rev. ed. London: Verso.

Anderson, B., and P. Harrison. 2006. "Questioning Affect and Emotion." *Area* 38:333–5.

Anderson, K., and S.J. Smith. 2001. "Editorial: Emotional Geographies." *Transactions of the Institute of British Geographers*, n.s., 26:7–10.

Arnold, D. 1996. *The Problem of Nature: Environment, Culture and European Expansion*. Oxford: Blackwell.

Arpin, I., and C. Granjou. 2015. "The Right Time for the Job? Insights into Practices of Time in Contemporary Field Sciences." *Science in Context* 28:237–58.

Atwood, M. 1972. *Survival: A Thematic Guide to Canadian Literature*. Toronto: House of Anansi.

– 1995. *Strange Things: The Malevolent North in Canadian Literature*. Oxford: Clarendon.

– 2003. "Survival Then and Now." In *The Canadian Distinctiveness into the XXIst Century / La distinction canadienne au tournant du XXIe siècle*, edited by C. Gaffield and K.L. Gould, 47–55. Ottawa: Les Presses de l'Université d'Ottawa / University of Ottawa Press.

Bakhtin, M. 1984. *Rabelais and His World*. Translated by H. Iswolsky. Bloomington: Indiana University Press.

Barbarito, M., S. Baldanza, and A. Peri. 2001. "Evolution of the Coping Strategies in an Isolated Group in an Antarctic Base." *Polar Record* 37 (201):111–20.

Barnes, T.J. 2001. "Retheorizing Economic Geography: From the Quantitative Revolution to the 'Cultural Turn.'" *Annals of the Association of American Geographers* 91:546–65.

Barnes, T.J., and M. Farish. 2006. "Between Regions: Science, Militarism and American Geography from World War to Cold War." *Annals of the Association of American Geographers* 96:807–26.

Barnett, C. 1998. "Impure and Worldly Geography: The Africanist Discourse of the Royal Geographical Society, 1831–73." *Transactions of the Institute of British Geographers*, n.s., 23:239–51.

BBC (British Broadcasting Corporation). 2000. *Vets in the Wild: Polar Bear Special*. London: BBC.

Beatty, A. 2005a. "Emotions in the Field: What Are We Talking About?" *Journal of the Royal Anthropological Institute* 11:17–37.

– 2005b. "Feeling Your Way in Java: An Essay on Society and Emotion." *Ethnos* 70:53–78.

Beeby, D. 1994. *In a Crystal Land: Canadian Explorers in Antarctica*. Toronto: University of Toronto Press.

Berger, C. 1966. "The True North Strong and Free." In *Nationalism in Canada*, edited by P. Russell, 3–26. Toronto: McGraw-Hill Canada.

Berry, M.J., and K.G. Barr. 1971. "A Seismic Refraction Profile across the Polar Continental Shelf of the Queen Elizabeth Islands." *Canadian Journal of Earth Science* 8:347–70.

Besnier, N. 1990. "Language and Affect." *Annual Review of Anthropology* 19: 419–51.

Bhabha, H.K. 1994. *The Location of Culture*. London: Routledge.

Boas, F. 1888. *The Central Eskimo*. Washington, DC: Bureau of Ethnology, Smithsonian Institution.

Bondi, L. 2003. "Empathy and Identification: Conceptual Resources for Feminist Fieldwork." *ACME: An International E-Journal for Critical Geographies* 2 (1): 64–76.

– 2005. "The Place of Emotions in Research: From Partitioning Emotion and Reason to the Emotional Dynamics of Research Relationships." In

Emotional Geographies, edited by J. Davidson, L. Bondi, and M. Smith, 231–46. Aldershot, UK: Ashgate.
Bondi, L., J. Davidson, and M. Smith. 2005. "Introduction: Geography's 'Emotional Turn.'" In *Emotional Geographies*, edited by J. Davidson, L. Bondi, and M. Smith, 1–16. Aldershot, UK: Ashgate.
Braun, B. 2000. "Producing Vertical Territory: Geology and Governmentality in Late Victorian Canada." *Ecumene* 7:7–46.
Bravo, M.T. 1999. "Ethnographic Navigation and the Geographical Gift." In *Geography and Enlightenment*, edited by D.N. Livingstone and C.W.J. Withers, 199–235. Chicago: University of Chicago Press.
– 2000. "The Rhetoric of Scientific Practice in Nunavut." *Ecumene* 7:468–74.
– 2002. "Measuring Danes and Eskimos." In *Narrating the Arctic: A Cultural History of Nordic Scientific Practices*, edited by M. Bravo and S. Sörlin, 235–73. Canton, MA: Science History Publications/USA.
Bravo, M.T., and S. Sörlin, eds. 2002. *Narrating the Arctic: A Cultural History of Nordic Scientific Practices*. Canton, MA: Science History Publications/USA.
Brennan, P.H. 1998. "Diefenbaker and the Press: A Case Study of *Maclean's*, 1958–1963." In *The Diefenbaker Legacy: Canadian Politics, Law and Society since 1957*, edited by D.C. Story and R.B. Shepard, 137–53. Regina: Canadian Plains Research Center, University of Regina.
Briggs, J.L. 1970. *Never in Anger: Portrait of an Eskimo Family*. Cambridge, MA: Harvard University Press.
– 1987. "In Search of Emotional Meaning." *Ethos* 15:8–15.
– 1998. *Inuit Morality Play: The Emotional Education of a Three-Year-Old*. New Haven, CT: Yale University Press.
Brody, H. 1975. *The People's Land: Eskimos and Whites in the Eastern Arctic*. Harmondsworth, UK: Penguin Books.
Burnett, D.G. 2000. *Masters of All They Surveyed: Exploration, Geography and a British El Dorado*. Chicago: University of Chicago Press.
Campbell, D.W. 1998. "Preface." In *Canada and the Early Cold War 1943–1957 / Le Canada au début de la guerre froide 1943–1957*, edited by G. Donaghy, 8–12. Ottawa: Department of Foreign Affairs and International Trade, Canada and Canadian Government Publishing.
Church, M. 1984. "On Experimental Method in Geomorphology." In *Catchment Experiments in Geomorphology*, edited by T.P. Burt and D.E. Walling, 563–80. Norwich, UK: Geobooks.
Church, M., B. Gomez, E.J. Hickin, and O. Slaymaker. 1985. "Geomorphological Sociology." *Earth Surface Processes and Landforms* 10:539–40.
Church, M., and D.M. Mark. 1980. "On Size and Scale in Geomorphology." *Progress in Physical Geography* 4:342–90.

Church, M., and O. Slaymaker. 1985. "Preface." In *Field and Theory: Lectures in Geocryology*, edited by M. Church and O. Slaymaker, xi. Vancouver: University of British Columbia Press.

Cloud, J. 2000. "Crossing the Olentangy River: The Figure of the Earth and the Military-Industrial-Academic-Complex, 1947–1972." *Studies in History and Philosophy of Modern Physics* 31:371–404.

– 2001a. "Hidden in Plain Sight: The CORONA Reconnaissance Satellite Programme and Clandestine Cold War Science." *Annals of Science* 58:203–9.

– 2001b. "Imaging the World in a Barrel: CORONA and the Clandestine Convergence of the Earth Sciences." *Social Studies of Science* 31:231–51.

– 2003. "Introduction: Special Guest-Edited Issue on the Earth Sciences in the Cold War." *Social Studies of Science* 33:629–33.

Coates, K.S., and W.R. Morrison. 1996. "Writing the North: A Survey of Contemporary Canadian Writing on Northern Regions." *Essays on Canadian Writing* 59:5–25.

Collins, H.M. 2001. "What Is Tacit Knowledge?" In *The Practice Turn in Contemporary Theory*, edited by T.R. Schatzki, K. Knorr Cetina, and E. von Savigny, 107–19. London: Routledge.

Collis, C. 1996. "The Voyage of the Episteme: Narrating the North." *Essays on Canadian Writing* 59:26–45.

– 2003. "Demythologizing Literature's North." *Essays on Canadian Writing* 79:155–62.

Cook, R. 2005. *Watching Quebec: Selected Essays*. Montreal and Kingston: McGill-Queen's University Press.

Corsín Jiménez, A. 2003. "Working Out Personhood: Notes on Labour and Its Anthropology." *Anthropology Today* 19 (5): 14–17.

Cox, D.R. 1958. *Planning of Experiments*. New York: John Wiley and Sons.

Craib, R.B. 2004. *Cartographic Mexico: A History of State Formations and Fugitive Landscapes*. Durham, NC: Duke University Press.

Cruikshank, J. 1990. *Life Lived Like a Story: Life Stories of Three Yukon Native Elders*. In collaboration with A. Sidney, K. Smith, and A. Ned. Vancouver: University of British Columbia Press.

– 1998. *The Social Life of Stories: Narrative and Knowledge in the Yukon Territory*. Vancouver: University of British Columbia Press.

– 2005. *Do Glaciers Listen? Local Knowledge, Colonial Encounters and Social Imagination*. Vancouver: University of British Columbia Press.

Daipha, P. 2015. *Masters of Uncertainty: Weather Forecasters and the Quest for Ground Truth*. Chicago: University of Chicago Press.

David, R.G. 2000. *The Arctic in the British Imagination 1818–1914*. Manchester: Manchester University Press.

de la Cadena, M., and M.E. Lien, eds. 2015. "Anthropology and STS: Generative Interfaces, Multiple Locations." *HAU: Journal of Ethnographic Theory* 5 (1): 437–75.
Dennis, M.A. 2003. "Earthly Matters: On the Cold War and the Earth Sciences." *Social Studies of Science* 33:809–19.
– 2006. "Secrecy and Science Revisited: From Politics to Historical Practice and Back." In *The Historiography of Contemporary Science, Technology, and Medicine: Writing Recent Science*, edited by R.E. Doel and T. Söderqvist, 172–84. London: Routledge.
Department of Fisheries and Oceans. 1993. *An Arctic Puzzle*. Ottawa: Department of Fisheries and Oceans, Government of Canada.
Dodds, K. 1997. "Antarctica and the Modern Geographical Imagination (1918–1960)." *Polar Record* 33 (184): 47–62.
– 2002. *Pink Ice: Britain and the South Atlantic Empire*. London: I.B. Tauris.
Doel, R.E. 2003. "Constituting the Postwar Earth Sciences: The Military's Influence on the Environmental Sciences in the USA after 1945." *Social Studies of Science* 33:635–66.
Doel, R.E., and K.C. Harper. 2006. "Prometheus Unleashed: Science as a Diplomatic Weapon in the Lyndon B. Johnson Administration." *Osiris* 21:66–85.
Doel, R.E., T.J. Levin, and M.K. Marker. 2006. "Extending Modern Cartography to the Ocean Depths: Military Patronage, Cold War Priorities, and the Heezen-Tharp Mapping Project, 1952–1959." *Journal of Historical Geography* 32:605–26.
Doing, P. 2004. "'Lab Hands' and the 'Scarlet O': Epistemic Politics and (Scientific) Labor." *Social Studies of Science* 34:299–323.
Donaghy, G. 1998. "Introduction." In *Canada and the Early Cold War 1943–1957 / Le Canada au début de la guerre froide 1943–1957*, edited by G. Donaghy, 14–30. Ottawa: Department of Foreign Affairs and International Trade, Canada and Canadian Government Publishing.
Drache, D. 1970. "The Canadian Bourgeoisie and Its National Consciousness." In *Close the 49th Parallel Etc.: The Americanization of Canada*, edited by I. Lumsden, 3–25. Toronto: University of Toronto Press.
Dyck, I. 1993. "Ethnography: A Feminist Method?" *Canadian Geographer / Le Géographe canadien* 37:52–7.
Edney, M.H. 1997. *Mapping an Empire: The Geographical Construction of British India, 1765–1843*. Chicago: University of Chicago Press.
Eldridge, C.C. 1997. "Introduction: The North Atlantic Triangle Revisited." In *Kith and Kin: Canada, Britain and the United States from Revolution to Cold War*, edited by C.C. Eldridge, xi–xxii. Cardiff: University of Wales Press.
Energy, Mines, and Resources Canada. 1986. *Islands in the Midnight Sun: The Story of the Polar Continental Shelf Project*. Toronto: Carleton.

England, J.H. 2000. "An Urgent Appeal to the Government of Canada to Proclaim Our Northern Identity." *Arctic* 53:204–9.
– 2010. "Canada Needs a Polar Policy." *Nature* 463 (7278): 159.
England, J.H., A.S. Dyke, and G.H.R. Henry. 1998. "Canada's Crisis in Arctic Science: The Urgent Need for an Arctic Science and Technology Policy; Or, 'Why Work in the Arctic? No One Lives There.'" *Arctic* 51:183–90.
Fischedick, K.S. 1995. *Practices and Pluralism: A Socio-Historical Analysis of Early Vegetation Science, 1900–1950*. Amsterdam: Centrale Drukkerj UvA.
Fisher, R.A. 1960. *The Design of Experiments*. 7th ed. Edinburgh: Oliver and Boyd.
Foster, M., and C. Marino. 1986. *The Polar Shelf: The Saga of Canada's Arctic Scientists*. Toronto: NC.
Frye, N. 1971. *The Bush Garden: Essays on the Canadian Imagination*. Toronto: House of Anansi.
Gable, E. 2002. "Beyond Belief? Play, Scepticism, and Religion in a West African Village." *Social Anthropology* 10:41–56.
Garfinkel, H. 1967. *Studies in Ethnomethodology*. Englewood Cliffs, NJ: Prentice Hall.
Geertz, C. 1973. *The Interpretation of Culture: Selected Essays*. New York: Basic Books.
– 2000. *Available Light: Anthropological Reflections on Philosophical Topics*. Princeton, NJ: Princeton University Press.
Gieryn, T.F. 1983. "Boundary-Work and the Demarcation of Science from Non-Science: Strains and Interests in Professional Ideologies of Scientists." *American Sociological Review* 48:781–95.
– 1999. *Cultural Boundaries of Science: Credibility on the Line*. Chicago: University of Chicago Press.
– 2008. "Laboratory Design for Post-Fordist Science." *Isis* 99:796–802.
Gloin, K.J. 1998. "Canada-US Relations in the Diefenbaker Era: Another Look." In *The Diefenbaker Legacy: Canadian Politics, Law and Society since 1957*, edited by D.C. Story and R.B. Shepard, 1–14. Regina: Canadian Plains Research Center, University of Regina.
Gooday, G. 2008. "Placing or Replacing the Laboratory in the History of Science?" *Isis* 99:783–95.
Gooding, D., T. Pinch, and S. Schaffer, eds. 1989. *The Uses of Experiment: Studies in the Natural Sciences*. Cambridge: Cambridge University Press.
Gould, G. (1967) 1985. "'The Idea of North': An Introduction." Reprinted in *The Glenn Gould Reader*, edited by T. Page, 391–94. New York: Alfred A. Knopf.
Grace, S. 1996. "Introduction: Representing North (or, Greetings from Nelvana)." *Essays on Canadian Writing* 59:1–4.

– 2001. *Canada and the Idea of North*. Montreal and Kingston: McGill-Queen's University Press.
Grant, S.D. 1988. *Sovereignty or Security? Government Policy in the Canadian North 1936–1950*. Vancouver: University of British Columbia Press.
Green, S.F. 2005. *Notes from the Balkans: Locating Marginality and Ambiguity on the Greek-Albanian Border*. Princeton, NJ: Princeton University Press.
Greene, M.T. 1989. "Afterword." In *From Hutton to Hack (Binghamton Symposia in Geomorphology: International Series, no. 19)*, edited by K.J. Tinkler, 325–31. Boston: Unwin Hyman.
Gregory, D. 2004. *The Colonial Present: Afghanistan, Palestine, Iraq*. Oxford: Blackwell.
Griffiths, R. 2000. "Introduction." In *Great Questions of Canada*, edited by R. Griffiths, ix–xiii. Toronto: Stoddart.
Gusterson, H. 1996. *Nuclear Rites: A Weapons Laboratory at the End of the Cold War*. Berkeley: University of California Press.
Hacking, I. 1983. *Representing and Intervening: Introductory Topics in the Philosophy of Natural Science*. Cambridge: Cambridge University Press.
Haliburton, R.G. 1869. *The Men of the North and Their Place in History: A Lecture Delivered before the Montreal Literary Club, March 31st, 1869*. Montreal: John Lovell.
Hamelin, L.-E. 1979. *Canadian Nordicity: It's Your North Too*. Translated by W. Barr. First published as *Nordicité canadienne* in 1978. Montreal: Harvest House.
– 1988a. *The Canadian North and Its Conceptual Referents*. Ottawa: About Canada / Canadian Studies Directorate, Department of the Secretary of State of Canada.
– 1988b. "Louis-Edmond Hamelin Receives Northern Science Award [speaking notes used by Hamelin when accepting the Centenary Medal for Northern Science in 1986]." *Musk-ox* 36:59–63.
Harré, R. 1981. *Great Scientific Experiments: Twenty Experiments That Changed the World*. Oxford: Phaidon.
Harris, C. 2001. "Postmodern Patriotism: Canadian Reflections." *Canadian Geographer / Le Géographe canadien* 45:193–207.
Harvey, D. 1969. *Explanation in Geography*. London: Edward Arnold.
Hayden, C. 2003. *When Nature Goes Public: The Making and Unmaking of Bioprospecting in Mexico*. Princeton, NJ: Princeton University Press.
Heffernan, M. 2001. "'A Dream as Frail as Those of Ancient Time': The In-Credible Geographies of Timbuctoo." *Environment and Planning D: Society and Space* 19:203–25.
Henighan, S. 2002. *When Words Deny the World: The Reshaping of Canadian Writing*. Erin, ON: Porcupine's Quill.

Henke, C.R. 2000. "Making a Place for Science: The Field Trial." *Social Studies of Science* 30:483–511.
Herbert, S. 2000. "For Ethnography." *Progress in Human Geography* 24:550–68.
– 2001. "From Spy to Okay Guy: Trust and Validity in Fieldwork with the Police." *Geographical Review* 91:304–10.
Hevly, B. 1996. "The Heroic Science of Glacier Motion." In *Science in the Field. Osiris* 2nd ser. 11, edited by H. Kuklick and R.E. Kohler, 66–86. Chicago: University of Chicago Press.
Hicks, J., and G. White. 2000. "Nunavut: Inuit Self-Determination through a Land Claim and Public Government?" In *Nunavut: Inuit Regain Control of Their Lands and Their Lives*, edited by J. Dahl, J. Hicks, and P. Jull, 30–115. Copenhagen: International Working Group for Indigenous Affairs.
Hill, J. 2008. *White Horizon: The Arctic in the Nineteenth-Century British Imagination*. Albany: State University of New York Press.
Hobson, G. 1990. "Polar Continental Shelf Project." In *Canada's Missing Dimension: Science and History in the Canadian Arctic Islands*, edited by C.R. Harington, 1:9–21. Ottawa: Canadian Museum of Nature.
Hochschild, A.R. 1983. *The Managed Heart: Commercialization of Human Feeling*. Berkeley: University of California Press.
House of Commons. 1958. *House of Commons Debates. Official Report Canada, Volume 4*. Ottawa: Government of Canada.
Hulan, R. 2002. *Northern Experience and the Myths of Canadian Culture*. Montreal and Kingston: McGill-Queen's University Press.
Innis, H.A. (1930) 1970. *The Fur Trade in Canada*. Rev. ed. Toronto: University of Toronto Press.
Ipellie, A. 1993. *Arctic Dreams and Nightmares*. Penticton, BC: Theytus Books.
– 1997. "Thirsty for Life: A Nomad Learns to Write and Draw." In *Echoing Silence: Essays on Arctic Narrative*, edited by J. Moss, 93–101. Ottawa: University of Ottawa Press.
Isinger, R., and D.C. Story. 1998. "The Plane Truth: The Avro Canada CF-105 Arrow Program." In *The Diefenbaker Legacy: Canadian Politics, Law and Society since 1957*, edited by D.C. Story and R.B. Shepard, 43–55. Regina: Canadian Plains Research Center, University of Regina.
Jackson, P. 1992. "The Politics of the Streets: A Geography of Caribana." *Political Geography* 11:130–51.
Jahn, A. 1985. "Experimental Observations of Periglacial Processes in the Arctic." In *Field and Theory: Lectures in Geocryology*, edited by M. Church and O. Slaymaker, 17–35. Vancouver: University of British Columbia Press.
Jones, M. 2003. *The Last Great Quest: Captain Scott's Antarctic Sacrifice*. Oxford: Oxford University Press.

Jones-Imhotep, E. 2000. "Disciplining Technology: Electronic Reliability, Cold-War Military Culture and the Topside Ionogram." *History and Technology* 17:125–75.
Josephides, L. 2005. "Resentment as a Sense of Self." In *Mixed Emotions: Anthropological Studies of Feeling*, edited by K. Milton and M. Svašek, 71–90. Oxford: Berg.
Kennedy, B.A. 1977. "A Question of Scale?" *Progress in Physical Geography* 1:154–7.
– 1979. "A Naughty World." *Transactions of the Institute of British Geographers*, n.s., 4:550–8.
– 1992. "Hutton to Horton: Views of Sequence, Progression and Equilibrium in Geomorphology." *Geomorphology* 5:231–50.
Kilbourn, W. 1970. *Pipeline: TransCanada and the Great Debate – A History of Business and Politics*. Toronto: Clarke, Irwin.
Kirsch, S. 1998. "Experiments in Progress: Edward Teller's Controversial Geographies." *Ecumene* 5:267–85.
Knorr Cetina, K. 1999. *Epistemic Cultures: How the Sciences Make Knowledge*. Cambridge, MA: Harvard University Press.
Kobalenko, J. 2002. *The Horizontal Everest: Extreme Journeys on Ellesmere Island*. Toronto: Viking/Penguin.
Kochan, J. 2015. "Objective Styles in Northern Field Science." *Studies in History and Philosophy of Science* 52:1–12.
Kohler, R.E. 2002a. *Landscapes and Labscapes: Exploring the Lab-Field Border in Biology*. Chicago: University of Chicago Press.
– 2002b. "Place and Practice in Field Biology." *History of Science* 40:189–210.
– 2008. "Lab History: Reflections." *Isis* 99:761–8.
Kuklick, H. 2011. "Personal Equations: Reflections on the History of Fieldwork, with Special Reference to Sociocultural Anthropology." *Isis* 102:1–33.
Kuklick, H., and R.E. Kohler. 1996. "Introduction." In *Science in the Field. Osiris* 2nd ser. 11, edited by H. Kuklick and R.E. Kohler, 1–14. Chicago: University of Chicago Press.
Kymlicka, W. 2001. *Politics in the Vernacular: Nationalism, Multiculturalism, and Citizenship*. Oxford: Oxford University Press.
– 2003. "Being Canadian." *Government and Opposition* 38:357–85.
LaRose, J. 2000. *Mars on Earth: Preparing for Life on the Red Planet*. Toronto: Summerhill Entertainment, Discovery Channel Canada, and Canadian Geographic.
Latour, B. 1990. "The Force and the Reason of Experiment." In *Experimental Inquiries: Historical, Philosophical and Social Studies of Experimentation in Science*, edited by H.E. Le Grand, 49–80. Dordrecht: Kluwer Academic Publishers.

– 1999. *Pandora's Hope: Essays on the Reality of Science Studies*. Cambridge, MA: Harvard University Press.

Latour, B., and S. Woolgar. 1986. *Laboratory Life: The Construction of Scientific Facts*. 2nd ed. Princeton, NJ: Princeton University Press.

Laugrand, F., and J. Oosten. 2002. "Inuit and Qallunaaq Perspectives: Interacting Points of View." *Études/Inuit/Studies* 26:13–15.

Lave, J. 1988. *Cognition in Practice: Mind, Mathematics and Culture in Everyday Life*. Cambridge: Cambridge University Press.

Lave, J., and E. Wenger. 1991. *Situated Learning: Legitimate Peripheral Participation*. Cambridge: Cambridge University Press.

Law, J. 1994. *Organizing Modernity*. Oxford: Blackwell.

Le Grand, H.E. 1990. "Is a Picture Worth a Thousand Experiments?" In *Experimental Inquiries: Historical, Philosophical and Social Studies of Experimentation in Science*, edited by H.E. Le Grand, 241–70. Dordrecht: Kluwer Academic Publishers.

Levy, I. 1988. *Search for a "Tropical Arctic."* Toronto: Royal Canadian Geographical Society, Breakthrough Films, Energy, Mines, and Resources Canada and Canadian Geographic.

Lewis, C., and S. Pile. 1996. "Woman, Body, Space: Rio Carnival and the Politics of Performance." *Gender, Place and Culture* 3:23–41.

Ley, D. 1974. *The Black Inner City as Frontier Outpost: Images and Behavior of a Philadelphia Neighborhood*. Monograph Series No. 7. Washington, DC: Association of American Geographers.

Livingstone, D.N. 2002. *Science, Space and Hermeneutics, Hettner-Lectures 5 (2001)*. Heidelberg: Department of Geography, University of Heidelberg.

– 2003. *Putting Science in Its Place: Geographies of Scientific Knowledge*. Chicago: University of Chicago Press.

Lorimer, H. 2003a. "The Geographical Field Course as Active Archive." *Cultural Geographies* 10:278–308.

– 2003b. "Telling Small Stories: Spaces of Knowledge and the Practice of Geography." *Transactions of the Institute of British Geographers*, n.s., 28:197–217.

Low, R. 1991. *Secrets of Ice*. Toronto: Breakthrough Films and Television and CKVR-TV.

Luhrmann, T.M. 2001. "The Touch of the Real." *Times Literary Supplement*, 12 January, 3–4.

Lury, C. 2003. "The Game of Loyalt(o)y: Diversions and Divisions in Network Society." *Sociological Review* 51:301–20.

MacDonald, F. 2006a. "Geopolitics and 'The Vision Thing': Regarding Britain and America's First Nuclear Missile." *Transactions of the Institute of British Geographers*, n.s., 31:53–71.

– 2006b. “The Last Outpost of Empire: Rockall and the Cold War.” *Journal of Historical Geography* 32:627–47.

MacLennan, H. (1945) 1967. *Two Solitudes*. Reprint. Toronto: Macmillan.

Marcus, A.R. 1995. *Relocating Eden: The Image and Politics of Inuit Exile in the Canadian Arctic*. Hanover, NH: University Press of New England.

Marcus, G.E. 1986. “Afterword: Ethnographic Writing and Anthropological Careers.” In *Writing Culture: The Poetics and Politics of Ethnography*, edited by J. Clifford and G.E. Marcus, 262–6. Berkeley: University of California Press.

– 1998. *Ethnography through Thick and Thin*. Princeton, NJ: Princeton University Press.

– 2000. “The Twistings and Turnings of Geography and Anthropology in Winds of Millennial Transition.” In *Cultural Turns / Geographical Turns: Perspectives on Cultural Geography*, edited by I. Cook, D. Crouch, S. Naylor, and J.R. Ryan, 13–25. Harlow, UK: Prentice Hall.

Masco, J. 2006. *The Nuclear Borderlands: The Manhattan Project in Post-Cold War New Mexico*. Princeton, NJ: Princeton University Press.

Mathews, W.H. 1985. “On the Scientific Method of J. Ross Mackay.” In *Field and Theory: Lectures in Geocryology*, edited by M. Church and O. Slaymaker, 1–16. Vancouver: University of British Columbia Press.

Matless, D. 2003. “Original Theories: Science and the Currency of the Local.” *Cultural Geographies* 10:354–78.

Mauss, M. 1906. “Essai sur les variations saisonnières des sociétés eskimo: Étude de morphologie sociale [avec la collaboration de H. Beuchat].” *L'Année sociologique* 9:39–132.

McCormack, D. 2006. “For the Love of Pipes and Cables: A Response to Deborah Thien.” *Area* 38:330–2.

McDowell, L. 1997. *Capital Culture: Gender at Work in the City*. Oxford: Blackwell.

McGhee, R. 2001. *Ancient People of the Arctic*. Vancouver: University of British Columbia Press.

Merton, R.K. (1942) 1973. “The Normative Structure of Science.” Reprinted in *The Sociology of Science: Theoretical and Empirical Investigations*, edited by N.W. Storer, 267–78. Chicago: University of Chicago Press.

Miller, C.A. 2004. “Climate Science and the Making of a Global Political Order.” In *States of Knowledge: The Co-production of Science and Social Order*, edited by S. Jasanoff, 46–66. London: Routledge.

Milton, K. 2005a. “Afterword.” In *Mixed Emotions: Anthropological Studies of Feeling*, edited by K. Milton and M. Svašek, 215–24. Oxford: Berg.

– 2005b. “Meanings, Feelings and Human Ecology.” In *Mixed Emotions: Anthropological Studies of Feeling*, edited by K. Milton and M. Svašek, 25–41. Oxford: Berg.

Mitcham, A. 1983. *The Northern Imagination: A Study of Northern Canadian Literature*. Moonbeam, ON: Penumbra.
Mocellin, J.S.P. 1988. "A Behavioural Study of Human Responses to the Arctic and Antarctic Environments." PhD diss., University of British Columbia.
Mocellin, J.S.P., and P. Suedfeld. 1991. "Voices from the Ice: Diaries of Polar Explorers." *Environment and Behavior* 23:704–22.
Moffett, S.E. 1907. "The Americanization of Canada." PhD diss., Columbia University.
Morton, D. 1997. *A Short History of Canada*. 3rd rev. ed. Toronto: McClelland & Stewart.
Moss, J. 1996. *Enduring Dreams: An Exploration of Arctic Landscape*. Concord, ON: House of Anansi.
– 1999. *The Paradox of Meaning: Cultural Poetics and Critical Fictions*. Winnipeg: Turnstone.
Munro, M. 2009. "High Costs Ground Scientists." *Edmonton Journal*, 5 May.
Nadasdy, P.E. 2008. "Wildlife as Renewable Resource: Competing Conceptions of Wildlife, Time, and Management in the Yukon." In *Timely Assets: The Politics of Resources and Their Temporalities*, edited by E.E. Ferry and M.E. Limbert, 75–106. Santa Fe, NM: School for Advanced Research Press.
Nader, L. 1969. "Up the Anthropologist: Perspectives Gained from Studying Up." In *Reinventing Anthropology*, edited by D. Hymes, 284–311. New York: Pantheon Books.
– 1996. "Introduction: Anthropological Inquiry into Boundaries, Power, and Knowledge." In *Naked Science: Anthropological Inquiry into Boundaries, Power, and Knowledge*, edited by L. Nader, 1–25. New York: Routledge.
Nassichuk, W.W. 1987. "Forty Years of Northern Non-Renewable Natural Resource Development." *Arctic* 40:274–84.
Nixon, P.G. 1987. "Bureaucracy and Innovation." *Canadian Public Administration / Administration publique du Canada* 30:280–98.
Onfray, M. 2002. *Esthétique du Pôle Nord: Stèles hyperboréennes*. Paris: Bernard Grasset.
Ophir, A., and S. Shapin. 1991. "The Place of Knowledge: A Methodological Survey." *Science in Context* 4:3–21.
Oreskes, N. 1996. "Objectivity or Heroism? On the Invisibility of Women in Science." In *Science in the Field. Osiris* 2nd ser. 11, edited by H. Kuklick and R.E. Kohler, 87–113. Chicago: University of Chicago Press.
– 1999. *The Rejection of Continental Drift: Theory and Method in American Earth Science*. New York: Oxford University Press.
– 2003. "A Context of Motivation: US Navy Oceanographic Research and the Discovery of Sea-Floor Hydrothermal Vents." *Social Studies of Science* 33:697–742.

Oreskes, N., and J.R. Fleming. 2000. "Why Geophysics?" *Studies in the History and Philosophy of Modern Physics* 31:253–7.
Ortner, S.B. 1978. *Sherpas through Their Rituals*. Cambridge: Cambridge University Press.
– 1999. *Life and Death on Mt Everest: Sherpas and Himalayan Mountaineering*. Princeton, NJ: Princeton University Press.
– 2006. *Anthropology and Social Theory: Culture, Power, and the Acting Subject*. Durham, NC: Duke University Press.
– 2010. "Access: Reflections on Studying Up in Hollywood." *Ethnography* 11:211–33.
Outram, D. 1996. "New Spaces in Natural History." In *Cultures of Natural History*, edited by N. Jardine, J.A. Secord, and E.C. Spary, 249–65. Cambridge: Cambridge University Press.
– 1999. "On Being Perseus: New Knowledge, Dislocation, and Enlightenment Exploration." In *Geography and Enlightenment*, edited by D.N. Livingstone and C.W.J. Withers, 281–94. Chicago: University of Chicago Press.
Paine, R. 1977. "Preface." In *The White Arctic: Anthropological Essays on Tutelage and Ethnicity*, edited by R. Paine, xi–xiii. Toronto: University of Toronto Press.
Pálsson, G. 2001. *Writing on Ice: The Ethnographic Notebooks of Vilhjalmur Stefansson*. Hanover, NH: Dartmouth College, University Press of New England.
– 2002. "Arcticality: Gender, Race, and Geography in the Writings of Vilhjalmur Stefansson." In *Narrating the Arctic: A Cultural History of Nordic Scientific Practices*, edited by M. Bravo and S. Sörlin, 275–309. Canton, MA: Science History Publications/USA.
Parry, B. 2004. *Trading the Genome: Investigating the Commodification of Bio-Information*. New York: Columbia University Press.
Petrone, P., ed. 1988. *Northern Voices: Inuit Writing in English*. Toronto: University of Toronto Press.
–, ed. 1991. *First People, First Voices*. Toronto: University of Toronto.
Pickering, A. 1995. *The Mangle of Practice: Time, Agency, and Science*. Chicago: University of Chicago Press.
Polar Continental Shelf Project. 1974. *Islands in the Midnight Sun: The Story of the Polar Continental Shelf Project*. Ottawa: Information Canada.
Popper, K. (1957) 2002. *The Poverty of Historicism*. Reprint. London: Routledge.
Powell, R.C. 2002. "The Sirens' Voices? Field Practices and Dialogue in Geography." *Area* 34:261–72.
– 2005. "Resolute Base." In *Encyclopedia of the Arctic*, edited by M. Nuttall, 3:1760–1. New York: Routledge.

– 2007a. "Geographies of Science: Histories, Localities, Practices, Futures." *Progress in Human Geography* 31:309–29.
– 2007b. "'The Rigours of an Arctic Experiment': The Precarious Authority of Field Practices in the Canadian High Arctic." *Environment and Planning A* 39:1794–1811.
– 2008a. "Becoming a Geographical Scientist: Oral Histories of Arctic Fieldwork." *Transactions of the Institute of British Geographers*, n.s., 33:548–65.
– 2008b. "Guest Editorial: Configuring an 'Arctic Commons'?" *Political Geography* 27:827–32.
– 2008c. "Science, Sovereignty and Nation: Canada and the Legacy of the International Geophysical Year, 1957–1958." *Journal of Historical Geography* 34:618–38.
– 2009a. "Canada Day in Resolute: Performance, Ritual and the Nation in an Inuit Community." In *High Places: Cultural Geographies of Mountains, Ice and Science*, edited by D. Cosgrove and V. della Dora, 178–95. London: I.B. Tauris.
– 2009b. "Learning from Spaces of Play: Recording Emotional Practices in High Arctic Environmental Sciences." In *Emotion, Place and Culture*, edited by M. Smith, J. Davidson, L. Cameron, and L. Bondi, 115–32. Farnham, UK: Ashgate.
– 2010. "Lines of Possession? The Anxious Constitution of a Polar Geopolitics." *Political Geography* 29:74–7.
– 2015. "The Study of Geography? Franz Boas and His Canonical Returns." *Journal of Historical Geography* 49:21–30.
Pullman, P. 1995. *Northern Lights*. London: Scholastic Children's Books.
Rabinow, P. 1996. *Essays on the Anthropology of Reason*. Princeton, NJ: Princeton University Press.
– 2003. *Anthropos Today: Reflections on Modern Equipment*. Princeton, NJ: Princeton University Press.
– 2008. *Marking Time: On the Anthropology of the Contemporary*. Princeton, NJ: Princeton University Press.
Rasmussen, K. 1998. "Bureaucrats and Politicians in the Diefenbaker Era." In *The Diefenbaker Legacy: Canadian Politics, Law and Society since 1957*, edited by D.C. Story and R.B. Shepard, 155–67. Regina: Canadian Plains Research Center, University of Regina.
Reed, J.C. 1958. "The United States Turns North." *Geographical Review* 48: 321–35.
Rees, A. 2001. "Practising Infanticide, Observing Narrative: Controversial Texts in a Field Science." *Social Studies of Science* 31:507–31.
Resnick, P. 1970. "Canadian Defence Policy and the American Empire." In *Close the 49th Parallel Etc.: The Americanization of Canada*, edited by I. Lumsden, 93–115. Toronto: University of Toronto Press.

Riches, D. 1977. "Neighbours in the 'Bush': White Cliques." In *The White Arctic: Anthropological Essays on Tutelage and Ethnicity*, edited by R. Paine, 166–88. Toronto: University of Toronto Press.
– 1990. "The Force of Tradition in Eskimology." In *Localizing Strategies: Regional Traditions of Ethnographic Writing*, edited by R. Farndon, 71–89. Edinburgh and Washington, DC: Scottish Academic Press and Smithsonian Institution Press.
Richler, M. 1989. *Solomon Gursky Was Here*. Toronto: Viking.
Robertson, G. 2000. *Memoirs of a Very Civil Servant: Mackenzie King to Pierre Trudeau*. Toronto: University of Toronto Press.
Robinson, M.F. 2006. *The Coldest Crucible: Arctic Exploration and American Culture*. Chicago: University of Chicago Press.
Roots, E.F. 1960. "Canadian Polar Continental Shelf Project, 1959." *Polar Record* 10 (66): 275–6.
– 1962. "Canadian Polar Continental Shelf Project, 1959–62." *Polar Record* 11 (72): 270–6.
– 1969. "The Role of Logistics in Northern Research." In *Proceedings of the Second National Northern Research Conference*, edited by J.J. Bond, 65–77. Edmonton: Boreal Institute, University of Alberta.
Rosen, M. 1985. "Breakfast at Spiro's: Dramaturgy and Dominance." *Journal of Management* 11 (2): 31–48.
– 1988. "You Asked For It: Christmas at the Bosses' Expense." *Journal of Management Studies* 25:463–80.
Said, E.W. 1978. *Orientalism: Western Conceptions of the Orient*. London: Routledge and Kegan Paul.
– 1993. *Culture and Imperialism*. New York: Vintage Books.
Sandlos, J. 2007. *Hunters at the Margin: Native People and Wildlife Conservation in the Northwest Territories*. Vancouver: University of British Columbia Press.
Sauer, C.O. 1956. "The Education of a Geographer." *Annals of the Association of American Geographers* 46:287–99.
Saul, J.R. 1997. *Reflections of a Siamese Twin: Canada at the End of the Twentieth Century*. Toronto: Penguin Books Canada.
– 2003. "The Inclusive Shape of Complexity." In *The Canadian Distinctiveness into the XXIst Century / La distinction canadienne au tournant du XXIe siècle*, edited by C. Gaffield and K.L. Gould, 13–27. Ottawa: Les Presses de l'Université d'Ottawa / University of Ottawa Press.
Serreze, M.C., and R.G. Barry. 2005. *The Arctic Climate System*. Cambridge: Cambridge University Press.
Shapin, S. 1994. *A Social History of Truth: Civility and Science in Seventeenth-Century England*. Chicago: University of Chicago Press.

Shapin, S., and S. Schaffer. 1985. *Leviathan and the Air-Pump: Hobbes, Boyle, and the Experimental Life*. Princeton, NJ: Princeton University Press.

Simpson, E. 1998. "New Ways of Thinking about Nuclear Weapons and Canada's Defence Policy." In *The Diefenbaker Legacy: Canadian Politics, Law and Society since 1957*, edited by D.C. Story and R.B. Shepard, 27–41. Regina: Canadian Plains Research Center, University of Regina.

Simpson, G.G. 1963. "Historical Science." In *The Fabric of Geology*, edited by C.C. Albritton, 24–48. Reading, MA: Addison-Wesley.

Smith, C., and J. Agar. 1998. "Introduction: Making Space for Science." In *Making Space for Science: Territorial Themes in the Shaping of Knowledge*, edited by C. Smith and J. Agar, 1–23. Houndsmill, UK: Macmillan.

Smith, R.W. 2000. "A Setting for the International Geophysical Year." In *Reconsidering Sputnik: Forty Years since the Soviet Satellite*, edited by R.D. Launius, J.M. Logsdon, and R.W. Smith, 119–24. Amsterdam: Harwood Academic Publishers.

Sontag, S., and C. Drew (with A.L. Drew). 1998. *Blind Man's Buff: The Untold Story of American Submarine Espionage*. New York: Public Affairs.

Sörlin, S. 2002. "Rituals and Resources of Natural History: The North and the Arctic in Swedish Scientific Nationalism." In *Narrating the Arctic: A Cultural History of Nordic Scientific Practices*, edited by M. Bravo and S. Sörlin, 73–122. Canton, MA: Science History Publications/USA.

Spufford, F. 1996. *I May Be Some Time: Ice and the English Imagination*. London: Faber and Faber.

Stairs, D. 1998. "Realists at Work: Canadian Policy Makers and the Politics of Transition from Hot War to Cold War." In *Canada and the Early Cold War 1943–1957 / Le Canada au début de la guerre froide 1943–1957*, edited by G. Donaghy, 91–116. Ottawa: Department of Foreign Affairs and International Trade, Canada, and Canadian Government Publishing.

Stallybrass, P., and A. White. 1986. *The Politics and Poetics of Transgression*. Ithaca, NY: Cornell University Press.

Steinberg, P.E. 2001. *The Social Construction of the Ocean*. Cambridge: Cambridge University Press.

Stern, P. 2006a. "From Area Studies to Cultural Studies to a Critical Inuit Studies." In *Critical Inuit Studies: An Anthology of Contemporary Arctic Ethnography*, edited by P. Stern and L. Stevenson, 253–66. Lincoln: University of Nebraska Press.

– 2006b. "Land Claims, Development, and the Pipeline to Citizenship." In *Critical Inuit Studies: An Anthology of Contemporary Arctic Ethnography*, edited by P. Stern and L. Stevenson, 105–18. Lincoln: University of Nebraska Press.

Stevenson, L. 2006. "Introduction." In *Critical Inuit Studies: An Anthology of Contemporary Arctic Ethnography*, edited by P. Stern and L. Stevenson, 1–22. Lincoln: University of Nebraska Press.
– 2014. *Life beside Itself: Imagining Care in the Canadian Arctic*. Oakland: University of California Press.
Stuart, R.C. 1994. "Continentalism Revisited: Recent Narratives on the History of Canadian-American Relations." *Diplomatic History* 18:405–14.
Stuckenberger, A.N. 2005. *Community at Play: Social and Religious Dynamics in the Modern Inuit Community of Qikiqtarjuat*. Amsterdam: Rozenberg Publishers.
Sullivan, W. 1961. *Assault on the Unknown: The International Geophysical Year*. London: Hodder and Stoughton.
Svašek, M. 2005. "Introduction: Emotions in Anthropology." In *Mixed Emotions: Anthropological Studies of Feeling*, edited by K. Milton and M. Svašek, 1–23. Oxford: Berg.
Taylor, C. 1991. *The Ethics of Authenticity*. Cambridge, MA: Harvard University Press.
Tester, F.J. 2005. "Resolute Bay." In *Encyclopedia of the Arctic*, edited by M. Nuttall, 3:1761–2. New York: Routledge.
Tester, F.J., and P. Kulchyski. 1994. *Tammarniit (Mistakes): Inuit Relocation in the Eastern Arctic 1939–63*. Vancouver: University of British Columbia Press.
Thien, D. 2005. "After or Beyond Feeling? A Consideration of Affect and Emotion in Geography." *Area* 37:450–6.
Thisted, K. 2002. "The Power to Represent: Intertextuality and Discourse in *Smilla's Sense of Snow*." In *Narrating the Arctic: A Cultural History of Nordic Scientific Practices*, edited by M. Bravo and S. Sörlin, 311–42. Canton, MA: Science History Publications/USA.
Thomson, D.W. 1999. "Foreword." In *Mapping a Northern Land: The Survey of Canada, 1947–1994*, edited by G. McGrath and L.M. Sebert, vii–viii. Montreal and Kingston: McGill-Queen's University Press.
Thrift, N. 2008. *Non-Representational Theory: Space, Politics, Affect*. Abingdon, UK: Routledge.
Tolia-Kelly, D.P. 2006. "Affect: An Ethnocentric Encounter? Exploring the 'Universalist' Imperative of Emotional/Affectual Geographies." *Area* 38: 213–17.
Tonkin, E. 2005. "Being There: Emotion and Imagination in Anthropologists' Encounters." In *Mixed Emotions: Anthropological Studies of Feeling*, edited by K. Milton and M. Svašek, 55–69. Oxford: Berg.
Traweek, S. 1996. "*Kokusaika*, *Gaiatsu*, and *Bachigai*: Japanese Physicists' Strategies for Moving into the International Political Economy of Science."

In *Naked Science: Anthropological Inquiry into Boundaries, Power, and Knowledge*, edited by L. Nader, 174–97. New York: Routledge.
Turner, V. 1974. *Dramas, Fields, and Metaphors: Symbolic Action in Human Society*. Ithaca, NY: Cornell University Press.
– 1982. *From Ritual to Theatre: The Human Seriousness of Play*. New York: Performing Arts Journal Publications.
– 1987. *The Anthropology of Performance*. New York: Performing Arts Journal Publications.
van Herk, A. 1990. *Places Far from Ellesmere: A Geografictione*. Red Deer, AB: Red Deer College Press.
– 1997. "Creating William Barentsz; Piloting North." In *Echoing Silence: Essays on Arctic Narrative*, edited by J. Moss, 79–92. Ottawa: University of Ottawa Press.
van Steenburgh, W.E., Y.O. Fortier, and R. Thornsteinsson. 1964. "Scientific Research in the Arctic." In *The Unbelievable Land: 29 Experts Bring Us Closer to the Arctic*, edited by I.N. Smith, 108–13. Ottawa: Queen's Printer, Government of Canada.
Wachowich, N. 1999. *Saqiyuq: Stories from the Lives of Three Inuit Women*. In collaboration with A. Agalakti Awa, R. Kaukjak Katsak, and S. Pikujak Katsak. Montreal and Kingston: McGill-Queen's University Press.
Wacquant, L. 2005. "Carnal Connections: On Embodiment, Apprenticeship, and Membership." *Qualitative Sociology* 28:445–74.
Watson, R.A. 1966. "Is Geology Different: A Critical Discussion of 'The Fabric of Geology.'" *Philosophy of Science* 33:172–85.
Whitaker, R., and G. Marcuse. 1994. *Cold War Canada: The Making of a National Insecurity State, 1945–1957*. Toronto: University of Toronto Press.
Widdowfield, R. 2000. "The Place of Emotions in Academic Research." *Area* 32:199–208.
Wiebe, R. 1989. *Playing Dead: A Contemplation concerning the Arctic*. Edmonton: NeWest Publishers.
Withers, C.W.J. 2001. *Geography, Science and National Identity: Scotland since 1520*. Cambridge: Cambridge University Press.
– 2009. "Place and the 'Spatial Turn' in Geography and in History." *Journal of the History of Ideas* 70:637–58.
Yates, F. 1970. *Experimental Design: Selected Papers of Frank Yates*. London: Charles Griffin.

INDEX

adventure, North as locale of: "amateur" science and, 107; Diefenbaker's vision of, 27, 51–3 (*see also* Diefenbaker, John); expeditionary and experimental space, 54–5, 59, 69, 103–5; experimentation and, 66–7, 69, 74–5; tourism, 11, 104–5, 113, 126, 140, 166
affect: social embedding of, 153. *See also* emotional ethnographies
alcohol, 120, 131, 156, 160, 209n27, 212n37
Alert, 129
Alexander, Ralph, 135–8
Alexandra Fiord, Ellesmere Island, 85
Alistair (driver, Inuit), 137, 142–3, 157
Amagoalik, John, 133
Anderson, Benedict, 28
anthropology and geography, 7; emotional and affectual, 150–2 (*see also* emotional ethnographies). *See also* ethnographies of science; geographic sciences, practices of
arcticality, 30
Arctic Bay, 174
Arctic climate system, 3
Arctic development (Canadian), xi–xii. *See also* Canadian sovereignty in the Arctic; Diefenbaker, John
Arctic ethnography, tradition of, 25, 153–4. *See also* fieldwork, Arctic; play, spaces of
Arctic Institute of North America, University of Calgary, 29
Arctic lands: as "natural laboratory," 62–3; spaces of encounter, 126; in Western imagination, 125. *See also* nordicité canadienne (Canadian nordicity)
Arctic sovereignty. *See* Canadian sovereignty in the Arctic
Arnold, David, 30
Arnold, Keith, 67
Atwood, Margaret, 31–2, 34, 37
Avro Arrow, 42, 198–9n7
Axel Heiberg Island, 85

Baker, Richard, 106, 107
Bakhtin, Mikhail, 128, 148
Barnes, Trevor, 75
Barry (stores manager), 92–4, 155, 157, 158–60
Barry, Roger: *The Arctic Climate System*, 3
base cooks. *See* PCSP cooks and kitchen staff
base managers. *See* PCSP base managers

Beatty, Andrew, 150, 152
Beechey Island, 102
Ben (pilot), 109
Berry and Barr article (1971), 71–4, 74–5
Bhabha, Homi, 31
bilingualism, 68
Boas, Franz, 7
Bondi, Liz, 161–2, 207nn2–3
botanical fieldwork, 112–14, 113f, 205n24. *See also* fieldwork, Arctic
boundary-work, 103–7
Bradley's (flight contractor), 106, 170, 204n16, 210n14
Briggs, Jean, 148, 207n3; *Never in Anger*, 153–4
British Antarctic Survey (BAS), 107
Brock Island, 71, 72f
Brody, Hugh, 14, 127
Brown, Christopher, 89
Burry, Bert, 72–3, 166

Camp 200, 66f, 71–2, 73f
Canada Day, Resolute Bay, 127, 141–9, 141f, 144f, 145f; social spaces of the Arctic, 24–5, 127–9, 140, 148–9
Canada First movement, 32
Canada-Nunavut Geoscience Office, 211n25
Canadian Broadcasting Corporation, 33, 52
Canadian Forces in the Arctic. *See* military and defence
Canadian High Arctic Research Station (CHARS), xii, 189
Canadian identity and nationalism: Canada Day in Resolute, 145–9 (*see also* Canada Day, Resolute Bay); colony to nation, 40; Diefenbaker's government (*see* Diefenbaker, John); Inuit role in, 133; as middle power, 39; nordicity, 28, 32–6, 37–8, 189; PCSP, role in, 28–9, 42, 68, 166–8, 171–3, 187; post–Second World War, 41, 42; scientific sovereignty (*see* Canadian sovereignty in the Arctic; scientific sovereignty)
Canadian Journal of Earth Science: Berry and Barr article (1971), 71–4
Canadian literature, 34–5, 198n1
Canadian Rangers, 168–9, 170, 210n10
Canadian Remote Sensing Centre, 108
Canadian sovereignty in the Arctic, 48–53, 58, 131, 147, 173, 188–9, 191. *See also* Canadian identity and nationalism; Polar Continental Shelf Project (PCSP scientific sovereignty
Canadian Space Agency, 106
carnal sociology, 162
carnival, theories of, 128–9, 147, 148; carnivalesque and play, 155, 156–60; play and hierarchy, 156–60. *See also* Canada Day, Resolute Bay
cartographic projects. *See* mapping and maps
Catherine (graduate student), 164–5
Cetina, Karin Knorr, 98
Charles (base manager), 89–90, 167, 203n6, 210n13, 213n43
Church, Michael, 60
Claire (student), 115
climate change. *See* global climate change
Cloud, John, 56
Cold War: Canadian diplomacy, 40, 198n2; effect on geographical sciences, 56–7; national rivalries, 38, 131
Collin, Arthur, 51–2
Collins, Harry, 19
Collis, Christy, 33, 66–7
colonialism: Canadian post-colonialism, 31–2; colonial present, 138, 140, 148–9; of field sciences, 30; in histories of polar regions, 29–30; indigenous lands, 33; neo-colonial

relationship with US, 41; PCSP and Inuit relationships, 138; resistance by Inuit, 126
communication technology. *See* technology
communities of practice: description of concept, 18–20. *See also* scientific practices
Computing Devices of Canada, 71–2
Comtois, Paul, 48
continentalism, North American, 40–1
continental shelf, 46, 50–1, 200n19; seismic refraction profile, 71–4, 75. *See also* Polar Continental Shelf Project (PCSP)
Cook, Ramsay, 41
cooks and kitchen staff. *See* PCSP cooks and kitchen staff
Co-op Hotel, 157, 197n5
Co-op store, 136, 138, 139, 143
Cornwallis Island, 115, 129, 183
cost-recovery rates, 14, 100–1, 102, 103, 108, 180. *See also* funding
Cox, D.R., 61–2
Craig (base staff), 94, 158
Cresswell Bay, Somerset Island, 109, 115, 118
cultures of labour, 25. *See also* PCSP, study of social and cultural lives of

Daipha, Phaedra, 3
Decca 6f Lambda hyperbolic, 65, 71–2, 203n22. *See also* survey and navigation systems
defence and military. *See* military and defence
Defence Research Board (Canada), 39, 47, 119
Denmark, 30
Dennis, Michael, 56
Department of National Defence (Canada), 120, 170, 189
description (ethnographic), 21–2. *See also* ethnographies of science
Devon Island, 106
Diefenbaker, John, 29, 39–41, 188, 189, 198–9n6, 198nn3–5; “A New Vision” and Northern Vision, 27, 41–2, 45–6, 51, 52–3, 198n7, 199n8; northern focus of government, 43
difference: interpretive ethnographies, 164–5; resisting assimilation of, 163–4
Distant Early Warning (DEW), 43–4, 172, 200n13
Dodds, Klaus, 57, 205n17
Dominic (student), 115–18
Donaghy, Greg, 40
dreaming and the Magnetic North Pole, 15
Driver, Rex, 119–21
dry laboratories, 92, 92f, 114, 205n25

Elizabeth (postdoctoral assistant), 112–14
Ellef Ringnes Island, 50f, 64, 72f, 203n20
Ellesmere Island, 103–5, 186; narratives of, 35–6; Operation Hazen, 39
Emma (cleaner, Inuit), 137, 143
emotional ethnographies: emotional practices, 152; ethics and guilt, 160–2; gender within emotional labour, 175; geographic studies of, 150–2; social embedding of affect, 153. *See also* geographic sciences, practices of
empathy and ethnography, 163–4. *See also* emotional ethnographies
employment. *See* PCSP non-science and seasonal employees
entertainment and distractions: satellite TV, 185; Sony Walkman, 121–2. *See also* technology

environmental science: experimental practice, 61–3; social worlds of, 6. *See also* field sciences and scientists; fieldwork, Arctic
ethics of ethnographic research: literature on, 207nn2–3; notes on research methodology, 193–4; reflexivity and, 22, 164, 192; spaces of play, 160–2, 209n27
ethnographies of science: history of, 6–8; interconnected practices, 18–20; interpretive ethnographies, 164–5, 192; multi-site studies, 9, 21–2; need for interpretation, 124, 192; new outsider, 205n24, 206n13; reflexivity, 22–3, 124, 164, 192; re-visioning, 20–3; social sciences supported by PCSP, 103, 105. *See also* emotional ethnographies; PCSP, study of social and cultural lives of; place and space in ethnography
Eureka, Ellesmere Island, 129; radio, 85. *See also* Polar Continental Shelf Project (PCSP)
experiment: philosophy of, 58–61, 73–5. *See also* field sciences and scientists; fieldwork, Arctic

family, absent from, 158–9, 166, 171, 175–7
field sciences and scientists, 6–7; activity performed by humans, 190–1; domesticity and labour divisions, 154; emotional encounters, 152, 162 (*see also* emotional ethnographies); experiment, philosophy of, 58–61, 73–5; field experimentation, 1960s, 61–3, 73–4, 192, 202n2; mangle of practice, 55, 61, 74
fieldwork, Arctic: achievement of experimental field methods, 70–1, 203n20, 212n40; activity performed by humans, 190–1, 192; adaptation of experimental methods, 55–8, 74–5, 181–3; description of botanical, 112–15; description of wetland science, 115–18; environmental immunity, 67; future of, 181, 192; golden age of, 201n33; a good iceman, 122, 205n17; good practice, good field person, 24, 28–9, 63–8, 121–3, 124, 176, 185, 205n29, 211n29; Inuit and, 181; mechanical skills, 68; research and methodology blurred, 70–1, 181–3; rigour of, 69–74, 183; seismic refraction profile example, 71–4; structured observational measurements, 73; unique difficulties of, 181–3, 212n37
First Air, 76, 89, 93, 104, 137, 165–6, 177, 210n14
First United Nations Conference on the Law of the Sea (UNCLOS I), 46, 200n20
fly-camps, 12, 81f, 82, 84, 118–21, 119f; radio schedules, 84–7
foreign scientists, 170, 172, 208n16; in cultural hierarchy, 102–3, 108, 186
Frank (pilot), 109
Freeman, Edward (base manager), 84, 86, 90, 99–100, 102–3, 106, 109, 137, 155, 156, 157, 159, 167, 169–74, 176, 210n9, 212n33
French Canada, 213n43; Diefenbaker's government, 42
Frisch, Tom, 201n33
Frobisher Bay. *See* Iqaluit
funding: effect on kinds of research, 9; funding and transport availability, 100, 102–3, 105, 106, 179–80, 189, 212n33; PCSP boundary-work, 103–7; perception of PCSP budget, 173–4, 179–80; process for gaining support from PCSP, 99–103, 210n16; social status, 14; working conditions, 118

Gary (base staff), 77, 94, 158–9
Gavin (climatologist), 94
Geertz, Clifford, 8, 16, 19–20, 22
gender, 174–7, 184–5, 211n27
geographic sciences, practices of: emotional geographies, 150–2, 162 (*see also* emotional ethnographies); place and space in, 55–8. *See also* anthropology and geography; field sciences and scientists; fieldwork, Arctic; place and space in ethnography
geography, discipline of: anthropology and, 7; engagement with experiment, 75 (*see also* experiment); ethnographies of practices, 18–20; ethnography and, 20–3; history of geosciences, 55–6, 75; measuring nordicity, 34–5; theories, 75. *See also* ethnographies of science; place and space in ethnography
geography in Canadian identity. *See* Canadian identity and nationalism; nordicité canadienne (Canadian nordicity)
Geological Survey of Canada (GSC), 46, 47, 108, 211n25
geomorphology, 183–4, 212n39
Geophysics Research Directorate, US Air Force, Cambridge Research Center, 45
geopolitics of the Arctic. *See* Canadian identity and nationalism; Canadian sovereignty in the Arctic; Cold War; scientific sovereignty
Gieryn, Thomas, 103
global climate change: Circumpolar Region context, 3–4; encounters between Inuit and scientists, 127
global positioning systems (GPS), 88, 170–1. *See also* survey and navigation systems
Gluckman, Max, 128
good field scientist identity, 24, 28–9, 63–8, 121–3, 124, 176, 185, 205n29, 211n29. *See also* fieldwork, Arctic
Gould, Glenn: "Idea of the North," 33
Grace, Sherrill, 33, 34
Greene, Mott, 56
Greenland, 30
Grise Fiord, 105, 174
Group of Seven artists, 33

Hacking, Ian, 75
Haliburton, Robert G.: *The Men of the North and Their Place in History*, 32–3
Hamelin, Louis-Edmond, 34–5
Hamilton, Alvin, 43, 45–6, 69
Harper, Stephen, xi–xii, 188
Harré, Rom, 59
Harvey, David, 60
Heffernan, Michael, 17
Heidi (PCSP director), 101–2
helicopters, 64, 65f, 68, 71, 72–3, 82, 83f, 169, 210n11; crashes, 165–6; mechanics, 91. *See also* transportation
Henighan, Stephen, 31–2
Herbert, Steve, 15, 20
hermeneutic theorizing, 75
Hevly, Bruce, 55
High Arctic, Canada. *See* Arctic *(various)*; Polar Continental Shelf Project (PCSP); *individual places*
High Arctic Weather Station, Environment Canada, 130
history of geosciences. *See* geography, discipline of
Hobson, George, 44, 52, 200n14
Hochschild, Arlie, 175
hotels in Resolute, 110–11, 157, 197n5, 203n5, 205n19, 208n16
Howe, C.D., 40
Hulan, R., 36
Hunt, F.P., 66f, 73

Hunt, Frank, 67
hydrologists, 115f–17f

identity: Scandinavian's northern, 30–1. *See also* Canadian identity and nationalism
the imaginary: Canadian geography, 32–6; Canadian imaginary, 28–9, 29–32; Canadian imperialism, 33; Canadian literature, 34–5; multi-sited imaginary, 21–2
Indian and Northern Affairs Canada, 168
indigenous peoples: Canadian identity and, 37; Traditional Ecological Knowledge Program (TEK), 100, 101; wildlife management, 202n3. *See also* Inuit
Innes, Stuart, 165
Innis, Harold, 32
intercontinental ballistic missiles (ICBMs), 44, 56
International Geophysical Year, 1957–58, 29, 38–9, 44–5, 48, 50–1, 53
interviewees and observed, 194–6, 197n1; identity of voices, 161, 177; interviewees' discussions of interviews, 184; politics of, 13. *See also* ethics of ethnographic research; PCSP, study of social and cultural lives of; *individual interviewees*
Inuit: anthropology of emotions and, 153–4; Canadian identity, 145–7, 148–9; celebration of Canada Day, 141–5, 148–9 (*see also* Canada Day, Resolute Bay); celebration of Nunuvut Day, 138–9; community of Resolute, studies of, 9; de-colonial resistance by, 126; employment, 16, 136–8, 157–8, 208n16; history of relocation, 130–2; map of community locations, 10f; Parks Canada employees, 121; relationship with Qallunaat (non-Inuit), 127, 139–40, 168–9; relationship with RCMP, 134–5; relationship with scientists, 180, 181, 212n35; research and Inuit knowledges, 100, 101, 133
Inukjuak, 130–1
Iqaluit: about, 206n2; accommodation, 89, 203n5; flights, 76, 82, 87, 89; PCSP and, 174, 211n25, 212n32; studies of, 14, 127, 158. *See also* Nunavut, Territory of
Isachsen, Ellef Ringnes Island, 50f, 64, 68, 129

James (scientist), 121–2, 143, 144–5
Joint Arctic Weather Stations (JAWS), 10f, 129–30
Jones, Alex, 15
Jones, Max, 57
Joseph (student), 120–1
Josephides, Lisette, 161–2
journalists, 78, 97, 100, 133, 164–5, 209n1, 209n3, 213n42
Julian (contract with Parks Canada), 120–1, 212n37

Kenn Borek, 108–9, 137, 139–40, 166, 170
Kennedy, Barbara, 61
Kigliqaqvik Rangers, 168–9, 170, 210n10
King, William Lyon Mackenzie, 32
Kirk (Parks Canada), 119–21
Kirsch, Scott, 58
Kobalenko, Jerry, 140; *The Horizontal Everest*, 103–5
Kohler, Robert, 17–18, 55, 62
Kristin (base staff), 95, 209n26
Kuklick, Henrika, 17
Kymlicka, Will, 36

labour. *See* PCSP non-science and seasonal employees

landscape as geography, 35; in identity, 37
Latour, Bruno, 8, 18, 22, 58
Lave, Jean, 18–19, 154
Law, John, 22, 77
Law of the Sea Conference in Geneva, 1958, 29
Lefebvre, Quentin, 103, 118, 122, 157
licensing, scientific, 135–6, 178–9, 206n7
Livingstone, David, 63
Lomonosov Ridge, 188
Louis St Laurent (Canadian ice-breaker), 173
Luhrmann, T.M., 20
Lundgaard, Leif, 122, 205n28

MacDonald, John A., 41
Mackay, J.R., 69, 183, 212n40
MacLennan, Hugh, 33
Makivik Corporation, 131–2
Manchester School of Social Anthropology, 128
mapping and maps: Canadian Arctic Islands, 10f, 205n25; Canadian identity, 28; Canadian places in text, 5f; Canadian sovereignty, 45–6, 69; of fly-camps in base office, 81f, 82, 84; ICBMs, 44; measuring nordicity, 34–5; PCSP Arctic Field Operations, 1959–1966, 47f; practices of surveying, 57
Marcus, George, 21
Mark, David, 60
Mars Analogue research, Devon Crater, 106
Mars Society, 106
Martin (scientist), 79
McDowell, Linda, 108–9
McGhee, Robert, 113
mealtimes, 87
Meighen Island, 67, 203n20
Melville Island, 90, 123, 178
Merris, Russell, 105, 204n11
Merton, Alan, 89, 123, 177, 184
Merton, Robert, 52–3, 201n35
methodology, 193–4. *See also* ethics of ethnographic research
military and defence: Arctic and national defence, 42–4, 168, 188–9, 198–9n7; Avro Arrow, 42, 198–9n7; integration of continental air defence, 43–4; relationship with PCSP, 120, 144, 168, 172; relationship with Resolute, 130–1, 144; relationship with science, 49–50, 56–7. *See also* Cold War
Milton, Kay, 151–2
Mitcham, Alison, 34
Mitchell, Isabel, 170
Moss, John, 35
Mould Bay, Prince Patrick Island, 72–3, 118, 129
multiculturalism, 36–7, 133, 189
multi-sited research imaginary (Marcus), 21. *See also* ethnographies of science

Nanisivik, 166
Narwhal Hotel, 110–11, 197n5, 205n19
NASA, 106–7, 204n16
nationalism. *See* Canadian identity and nationalism
Newfoundlanders at PCSP: background to, 203n7; Ice Island, 208n17; mechanical and stores staff, 94, 213n43; in PCSP hierarchy, 155–60; seasonal employment, 136–7, 155–60. *See also* PCSP non-science and seasonal employees
New National Policy, 1958, 43, 53
Nicola (government scientist), 79, 112–14, 122, 174
nordicité canadienne (Canadian nordicity), 28, 32–6, 37–8, 189. *See also* Canadian identity and

nationalism; Canadian sovereignty in the Arctic
North American Air Defence Command (NORAD), 43
Northern Vision. *See* Diefenbaker, John
NSERC Northern Research Chair, 107
Nunavut, Territory of: celebration of Nunuvut Day, 138–9; creation of, 6, 125–6; Inuit identities, 133–4, 148; licensing procedures, 135–6; relationship with PCSP, 174, 177–8, 211n31. *See also* Iqaluit
Nunavut Land Claims Agreement (1993), 133–4
Nunavut Research Institute, 135–6, 147, 178, 212n32

Onfray, Michel: *Esthétique du Pôle Nord,* 3
Operation Franklin, Queen Elizabeth Islands, 64
Operation Hazen, 39
Oreskes, Naomi, 56
Ortner, Sherry, 25, 126
Outram, Dorinda, 17

Paine, Robert, 127
Pálsson, Gísli, 30–1
Parks Canada, 118–21, 119f
PCSP. *See* PCSP, study of social and cultural lives of; Polar Continental Shelf Project
Pearson, Lester B., 40, 198n5
Peter (pilot), 165–6
petroleum exploration, 46
Pickering, Andrew, 55
Pierre (Parks Canada), 119–21
place and space in ethnography: multi-sited imaginary, 21–2; practices of geographical sciences, 55–8; practices of place, 62; relationality of emotions, 151; spatial experience, 17–18. *See also* ethnographies of science; geographic sciences, practices of
play, spaces of: in Arctic ethnographic tradition, 153–4; emotional practices, 152–3; ethics of researching, 160–2; hierarchy and, 156–60. *See also* PCSP cultural hierarchy
Plummer, Michael, 183
polar bears, 111–15, 123, 137–8, 182–3
Polar Continental Shelf Project (PCSP): addressing Canadian sovereignty, 48–53, 69 (*see also* Canadian identity and nationalism; scientific sovereignty); Arctic Field Operations, 1959–1966, 47f; boundaries around Arctic field science, 103–7; Canadian public and, 67, 69, 133, 168, 172–3, 211n23; celebration of Canada Day, 142–5; description of, 5–6; establishing "good practice," 63–8, 202n8 (*see also* fieldwork, Arctic); historians of, 197n3; ice patrol, 73f; interdisciplinary team, 46, 53, 63–6, 74; logistical precedents set, 66–9; observational-experimental distinction, 58; origins of, 23–4, 42, 43–8, 200n24; PCSP Arctic-Antarctic Exchange Program, 101, 170; PCSP Ice Island, 158–9, 208n17; PCSP Steering Committee and Scientific Screening Committee, 185; relationship with federal government, 168–9, 172–3, 180–1; relationship with Resolute community, 135–40; renamed "Program," 189, 197n2; role of field experiments in culture of, 75; scope of activities, 82; survey party, 50f; synchronize scientific research in the Arctic, 48. *See also* fieldwork, Arctic; PCSP Resolute; PCSP, study of social and cultural lives of

- base managers: employment frustrations, 169–74; hierarchy, 90; multi-skilled, 90; role in base culture, 167–9, 186; scheduling and timing responsibilities, 87–90; working conditions, 88, 90
- cooks and kitchen staff: cultural study and, 78; food quality, 95–6, 110–11, 158; hierarchy and, 107–8, 157; introductions at orientations, 77; Inuit, 137; multi-skilled and as general resources, 68, 114–15, 156–7, 191; uncertain employment, 160; working hours and pay, 96, 156, 159
- cultural hierarchy, 13–14, 74–5; dry lab allocations, 114; gender, 175–7, 211n27; government employment, 169–71; initial navigation of, 77–8; maintenance of protocols, 86; naming of fly-camps, 82; Newfoundlanders in, 155–60; pilots, 108–10, 155; play and dissent, 153, 154, 155–60, 213n41; of scientists on base, 75, 86, 107–8, 112, 114, 185–7, 205n23, 209n2; social dynamics, 110–11, 116–18; transportation, 107–8, 112, 114, 179–80. *See also* power relations
- institutional rules: boundaries around Arctic field science, 103–7; to govern daily activities, 79, 124; managers' role in, 167–9; new rules introduced, 110–11; pilots and, 108–10; process for gaining support from PCSP 99–103; temperance, 98, 111, 123, 124
- non-science and seasonal employees: contribution to scientific work, 191; hierarchy among, 90, 157–8; Inuit, 136–8, 157–8; mechanics, 91, 139–40, 157; multi-skilled expectations, 68, 90; relationship with, beginning fieldwork, 77–8; staff from Newfoundland, 155–60 (*see also* Newfoundlanders); stores management, 92–3; studying down, 15–16; uncertain employment, 157, 159–60, 197n4, 208n19
- PCSP Resolute: good field person identity, 24; main building, 4f
- PCSP, study of social and cultural lives of: arrival and culture shock, 76–7; challenges of studying fieldwork, 16–18; gender relations, 174–7, 211n27; multi-site study, 8–12; observer and the observed, relationship, 77–8, 164; overview, 3–4, 6–8; performance of the everyday, 154–5; studying up and studying down, 12–16; technology and collegiality, 87. *See also* ethnographies of science; fieldwork, Arctic; interviewees and observed
- scientists: accommodation, 94–7, 95f, 96f; camaraderie, 184–6; Canada Day celebration, 142–4, 145, 148; cultural hierarchy of, 75, 86, 107–8, 112, 114, 185–7, 205n23, 209n2; discussions amongst, 177–9; funding discussions, 179–80; gender relations, 175–7; known as "beakers," 82; news broadcasts, 94; opinions of fieldwork, 181–4; relationship with cultural study of, 77–9; relationship with Inuit, 138, 139; relationship with PCSP, 179–81
- spatial organization: airport and buildings' relationship, 76–7; base office, 79–84, 80f, 81f; hangar, 91–4, 91f, 92f; main building (beaker), 94–7, 95f, 96f; radio schedules for fly-camps, 84–7; scheduling and timing, 87–90

polar histories, 29–32
Pond Inlet, 168–9, 174
Popper, Karl, 59

post-humanist theories of practice, 19–20
power relations: Canada Day inversion of, 128–9, 141–5, 148–9; research on affect, 151. *See also* PCSP cultural hierarchy
practices of geographic sciences. *See* geographic sciences, practices of
Prince Patrick Island, 71, 72f, 174
Pullman, Philip: *Northern Lights,* 54, 76

Qallunaat (non-Inuit): meaning of, 205n1
Qausuittuq, 131. *See also* Resolute Bay, Cornwallis Island, Nunavut
Quttinirpaaq Parks Canada base, Tanquary Fiord, Ellesmere Island, 118–21, 119f

Rabinow, Paul, 6
Rachel (cleaner, Inuit), 137, 139, 157
racism, 121, 126–7, 134, 140, 148–9, 155
Ramsay, Malcolm, 163, 165
reflexivity. *See under* ethnographies of science
research ethics. *See* ethics of ethnographic research
Resolute Bay, Cornwallis Island, Nunavut: airport, 11f, 76, 203n1; employment, 157–8, 208n16 (*see also* Inuit); ethnographic study base, 11–12; federal presences in, 134–40; hamlet of Resolute Bay, 131–2, 132f, 174; history as communications hub, 129–30; history of Inuit relocation, 130–2; unsupported visitors, 78. *See also* Canada Day, Resolute Bay; PCSP Resolute; PCSP, study of social and cultural lives of; Polar Continental Shelf Project (PCSP)
Riches, David, 153
Richler, Mordecai: *Solomon Gursky Was Here,* 31
Rigby, Bruce, 147, 212n32
ritual, theories of, 128–9, 148. *See also* Canada Day, Resolute Bay
Roads to Resources initiative, 43
Roger (base staff), 94
Roots, E.F. (Fred), 47, 49, 63f, 73, 183, 189, 202nn5–6, 208n19; establishing "good practice," 63–8, 69–70, 74, 122
Rose, Gillian, 184
Rosen, Michael, 209n27
Royal Canadian Mounted Police (RCMP), 130–1, 132, 134–5, 143, 155, 168–9, 204n11

Said, Edward, 28, 30
Sanagak Lake, 86
Sarah (base cook, Inuit), 78, 95, 137
satellite phones, 85–7, 171, 210n9
Saul, John Ralston, 37, 133
Scandinavian identities, 30–1
science as public service, xii, 48–53, 201n35
scientific licensing, 135–6, 178–9, 206n7
scientific practices: performance of, 19–20. *See also* emotional ethnographies
Scientific Screening Committee, PCSP, 99–102
scientific sovereignty, xii, 23–4, 38–9, 48–53, 189, 191. *See also* Canadian identity and nationalism; Polar Continental Shelf Project (PCSP)
"scientist-adventurer," 69–70
Scott Polar Research Institute (SPRI), 29, 64, 79
seismologists, 71
Serreze, Mark: *The Arctic Climate System,* 3
Shapin, Steven, 16

Smith, David, 102
smoking and cigarettes, 88
social and cultural lives of climate scientists. *See* emotional ethnographies; PCSP, study of social and cultural lives of
social dramas, 128–9
social sciences as "real" science, 103, 105, 204n9
Sörlin, Sverker, 30, 67
South Camp Inn, 197n5
space and place. *See* place and space in ethnography; play, spaces of
Sputnik, 38, 43, 53, 198–9n7
Stenkul Fiord, Ellesmere Island, 118
Stern, Pamela, 125
Stokes, George Gabriel, 44
St-Onge, Denis, 50–1, 52, 67, 68
storage and stores management, 92–3
Stuart (pilot), 166
Stuckenberger, Nicole, 154
studying up and studying down, 12–16
submarines, 44, 56
survey and navigation systems: Decca 6f Lambda hyperbolic, 65, 71–2, 203n22; global positioning systems (GPS), 88, 170–1
Svašek, M., 152
Sweden and northern identity, 30

Tanquary Fiord, Ellesmere Island, 85, 119f
technology: communication, 84–7, 95, 171, 176–7, 204n8, 210n9; contact ritual and, 143; global positioning systems (GPS), 88, 170–1; Internet use, 103, 106–7, 204n7. *See also* survey and navigation systems
Tellurometer, 66f
temperance (culture of), 98, 111, 120, 123, 124, 189–91. *See also* PCSP, study of social and cultural lives of
Tennyson, Lord Alfred, 41–2, 198n5
Territory of Nunavut. *See* Nunavut, Territory of
Terry (student), 119, 121
Thien, D., 151
Thisted, Kirsten, 30
Thomas, Mary Ellen, 147
Thule site, National Heritage site, 113
Tolia-Kelly, D.P., 151
Traditional Ecological Knowledge Program (TEK): funding, 100, 101
transgression, Arctic as area of, 35
transportation: all-terrain vehicles (ATVs), 86, 90–2, 94, 99, 106, 109, 115–17, 117f, 123, 203n3; complex support for scientists, 64, 65f, 68, 71, 72–3; cost of, 177; cultural hierarchy and, 107–8, 112, 114, 185; flight information, 87, 88–90; flights in and out, 76–7 (*see also* First Air; Kenn Borek); funding and transport availability, 100, 102–3, 105, 106, 179–80, 189, 212n33; hangar, 91–4, 91f, 92f; pilots, 108–10, 155, 165–6, 175, 179; risks of, 165–6; safety of, 170–1; shipping and packaging, 93; Skidoos, 91, 92, 94, 210n10; Twin Otters, 82, 83f, 87, 91, 121, 155, 166, 179, 210n11. *See also* helicopters
Trevor Bennet, 157
Trudeau, Justin, xi–xii
Truro Island, 118, 184
Tuktoyaktuk, Mackenzie Delta, 82, 157, 208n19, 209n26. *See also* Polar Continental Shelf Project (PCSP)
Turner, Victor, 25, 127–9, 147
Tyler, Steven, 103

UNCLOS I (First United Nations Conference on the Law of the Sea), 46, 200n20
UN Conference on the Law of the Sea, 1958, 39

UN Convention on the Law of the Sea, 188
United States: Arctic sovereignty, 43–4, 131, 147; joint weather stations, 129–30
USS *Nautilus,* 44, 200n14
USSR: Arctic sovereignty, 43, 131, 147

van Herk, Aritha, 35–6
van Steenburgh, W.E., 47, 48–9, 66
Vets in the Wild (BBC TV), 163, 164–5

Wacquant, Loïc, 162
Walder, Andrew (base manager), 79, 82, 84, 85, 89–90, 91, 100, 103–6, 122, 139, 159, 167–9, 176, 182–3, 213n43
Ward Hunt Island, 45f, 120–1
Warner, Cheryl, 15, 212n39
Water Survey Canada, 120
weather information, 85–6, 88–9
Wenger, Etienne, 18–19
wetland science, 115–18, 115f–17f
Widdowfield, Rebekah, 161
Wiebe, Rudy, 35
Winter Harbour, Melville Island, 129
Woolgar, S., 22

Yvonne (base manager), 82, 86–90, 155, 181–3, 210n13, 213n43